Quantum Mechanical Tunneling in Chemical Physics

Quantum Mechanical Tunneling in Chemical Physics

Hiroki Nakamura
Gennady Mil'nikov

CRC Press
Taylor & Francis Group
Boca Raton London New York

CRC Press is an imprint of the
Taylor & Francis Group, an **informa** business

CRC Press
Taylor & Francis Group
6000 Broken Sound Parkway NW, Suite 300
Boca Raton, FL 33487-2742

Version Date: 20121218

International Standard Book Number: 978-1-4665-0731-9 (Hardback)

Library of Congress Cataloging-in-Publication Data

Nakamura, Hiroki.
Quantum mechanical tunneling in chemical physics / Hiroki Nakamura, Institute of Molecular Science, National Chiao Tung University ... Taiwan, and Institute for Molecular Science, National Institutes of Natural Sciences, Okazaki, Japan, Gennady Mil'nikov, Faculty of Electrical, Electronic and Informatics Engineering, Osaka University ... Japan.
pages cm
Includes bibliographical references and index.
ISBN 978-1-4665-0731-9 (hardback)
1. Tunneling (Physics) I. Title.

QC176.8.T8N34 2013
537.6'226--dc23 2012047936

Visit the Taylor & Francis Web site at
http://www.taylorandfrancis.com

and the CRC Press Web site at
http://www.crcpress.com

Contents

Preface

In this book we mainly discuss the following three tunneling phenomena based on our recent theoretical developments: (1) energy splitting, or tunneling splitting, in symmetric double well potential, (2) decay of metastable state through tunneling, and (3) tunneling effects in chemical reactions. The corresponding physical processes can be naively understood in terms of a reaction path in which (1) two endpoints are fixed, (2) one endpoint is fixed and one is free, and (3) both ends are free. Quantum mechanical tunneling manifests itself in a very wide range of natural phenomena and the subjects we deal with here constitute only a small portion of the whole area. In spite of the long history of studies of quantum mechanical tunneling phenomena, however, the theory of multidimensional tunneling has not yet been satisfactorily developed. Recently, we have successfully developed practical and useful methods applicable to real multidimensional systems for the above-mentioned problems. Basic theories, practical methodologies, and actual numerical applications to real molecular systems are presented so that the reader can hopefully comprehend the basic concepts and dynamics of multidimensional tunneling phenomena and use the methods directly in various problems of molecular spectroscopy and chemical dynamics, if necessary. Insufficiency of the low-dimensional treatments that are often employed for practical problems and intriguing effects of multidimensionality are demonstrated and clarified conceptually as well as numerically. Furthermore, attention is called to the nonadiabatic tunneling phenomenon, which is quite ubiquitous in molecular systems and yet manifests unique and intriguing phenomenon in comparison with ordinary tunneling. In this book, quite a bit of mathematics is used for the development and explanation of basic theories; thus a background knowledge of mathematics is required at the level of graduate students. This book is intended as a standard reference for comprehending the phenomena and solving practical problems in the fields of chemical physics. H. N. mainly wrote the whole contents and thus is responsible for all chapters. G. M. shares responsibility for Chapters 6, 7, and 8 and Section 9.1.

We would like to thank all the collaborators for their contributions in the development of the methods and various applications to real molecular systems that clearly demonstrate the usefulness of our methods. H. N. would like to express his sincere thanks to National Chiao Tung University in Hsinchu, Taiwan, and people at the Institute of Molecular Science, Department of Applied Chemistry, Faculty of Science of the university for providing this nice opportunity to work on this book. Especially, he is deeply indebted to Professors S. H. Lin and Y. P. Lee for their hospitality. G. M. thanks his colleagues and friends in Okazaki who have turned this place into his second hometown after spending seven wonderful years in the Institute for Molecular Science, National Institutes of Natural Sciences. Acknowledgment is also due to the following societies and publishers for permission of reproducing various copyright materials: American Institute of Physics, American Chemical Society, Elsevier Science B.V., and World Scientific Publishing.

Finally, H. N. deeply thanks his wife, Suwako, for her continuous support in his life. Without her support in Taiwan, this book could not have been finished.

Hiroki Nakamura,
Hsinchu, Taiwan and Okazaki, Japan
and Gennady Mil'nikov,
Suita, Osaka, Japan,
September 2012

1 Introduction

Needless to say, quantum mechanical tunneling is the most fundamental and well-known quantum mechanical effect and has been a long-standing important subject since the dawn of quantum mechanics. It plays crucial roles in a wide range of natural sciences from various fields of physics and chemistry to biology, representing its multidisciplinarity. For instance, tunneling effects are studied in false vacuum states in quantum field theory [1–3], fission of atomic nuclei in nuclear physics [2,4,5], scanning tunneling microscopy and mesoscopic devices in solid-state physics [6–8], and proton transfer and chemical reactions in chemistry and biology [9–20]. The de Broglie wavelength λ of atoms and molecules is given by

$$\lambda = \frac{2\pi\hbar}{mv}, \tag{1.1}$$

where m and v are the mass and velocity of a particle. If a hydrogen atom moves with the kinetic energy of 1 eV, its corresponding de Broglie wavelength is as short as $\lambda \simeq 2.9 \times 10^{-9}$ cm and definitely its quantum mechanical behavior becomes important. At lower energies, the effect becomes more crucial.

There are enormous amounts of literature dealing with quantum mechanical tunneling phenomena, and many monographs have also been published recently covering various fields of the phenomenon [21–24]. Unfortunately, however, theories of multidimensional tunneling phenomena are not yet satisfactorily developed. Effective numerical algorithms in quantum mechanics such as MCTDH (multichannel time-dependent Hartree) method [25] and quantum Monte Carlo method [26] have been developed and successfully implemented to study quantum dynamics in multidimensional systems. However, exact quantum dynamics methods require a formidable task including accurate *ab initio* calculations of global potential energy surfaces. Besides, it is not straightforward to obtain the physical insights of dynamic processes.

In this sense, semiclassical methods seem to be the most promising, since any WKB (Wentzel-Kramers-Brillouin) type theory is based on classical trajectories that can be computed rather easily even in high-dimensional systems. So far, many semiclassical theories of multidimensional tunneling have been proposed and tested [1,4,7,27–37]. In the ordinary WKB-type approach, the crucial problem is matching the semiclassical wave functions at the boundary between the classically allowed region and the tunneling (classically forbidden) region in multidimensional space. This is equivalent to the analytical continuation into the complex phase space and has been done successfully so far only in two dimensions. The instanton approach, on the other hand, avoids this problem of matching by formally treating the energy E as the first-order term with respect to $\hbar$ in the semiclassical asymptotic expansion. As a result, in the limit of $\hbar \to 0$ the classically allowed region shrinks to a single point at the potential minimum position. The tunneling (instanton) trajectory has zero energy and originates from the minimum. This instanton theory was reformulated on the basis of the WKB approximation to the Schrödinger equation by treating the energy E as the first-order term with respect to $\hbar$ [33,34]. The main idea of instanton approach closely resembles that of periodic orbit method [27] and was originally formulated in the context of the path-integral formalism by Coleman [38] for the decay problem and by Vainshtein et al. [39] for the tunneling splitting. The decay problem means the decay of a metastable state supported by a potential well by tunneling through a potential barrier. The tunneling splitting is an energy splitting due to tunneling in a symmetric double well potential. This instanton approach was applied to relatively big molecules and molecular complexes [40–42]. There are, however, some problems in these treatments from our viewpoint. The contributions of the transverse modes perpendicular to the instanton orbit are assumed to be additive and the multidimensional systems are treated essentially as a set of two-dimensional ones. Another problem is related to the difficulty of finding the multidimensional instanton trajectory. It is not possible to find it just by shooting classical trajectories on the inverted potential energy surface, as can be done in two dimensions. Thus the instanton trajectory is found only approximately and the pre-exponential factor is often taken from one-dimensional theory.

Recently, we developed practical and useful semiclassical methods applicable to multidimensional systems [43–46]. These are multidimensional theories of tunneling splitting in symmetric double well potential and decay of metastable states through potential barrier. These are canonically invariant WKB-type instanton theory and based on exact solutions of Hamilton-Jacobi and transport equations (or continuity equation) for arbitrary multidimensional systems with general form of the Hamiltonian. The final result is presented in canonically invariant form, which does not resort to any special coordinate system and can be used directly for practical computations. The instanton trajectory can be determined efficiently in any multidimensional systems. We also formulated an efficient method to detect caustics, namely, a boundary between classically allowed and forbidden regions, on multidimensional potential energy surface [47]. Finding caustics is nontrivial in multidimensional space. The present method can be applied to tunneling in multidimensional chemical reactions, since tunneling trajectories can be run from the caustics on the inverted potential energy surface. These methods can be usefully applied to various realistic problems in molecular spectroscopy and chemical dynamics. Considering the boundary conditions of tunneling processes, there are three kinds of tunneling, as shown in Figure 1.1: (a) tunneling splitting in which both ends are bounded, (b) decay of metastable state through tunneling or resonance scattering via potential barrier penetration in which one end is free to infinity, and (c) tunneling in chemical reactions in which both ends are free. The above-mentioned theoretical methods are relevant to each of these subjects.

There is another important tunneling phenomenon that is called *nonadiabatic tunneling*. This represents one class of nonadiabatic transitions and plays significant roles in chemical dynamics [48]. The potential barrier is created by the interaction between two crossing diabatic potential energy surfaces with opposite signs of slopes (see Figure 1.2). This nonadiabatic tunneling is affected by the nonadiabatic coupling between the two adiabatic potentials and is quite different from ordinary tunneling. This phenomenon is rather ubiquitous in chemical physics, since the so-called conical intersections of potential energy surfaces appear frequently in molecular systems. This new type of tunneling requires careful treatment, but this is not necessarily well recognized yet. Tunneling through the lower adiabatic potential at energies lower than

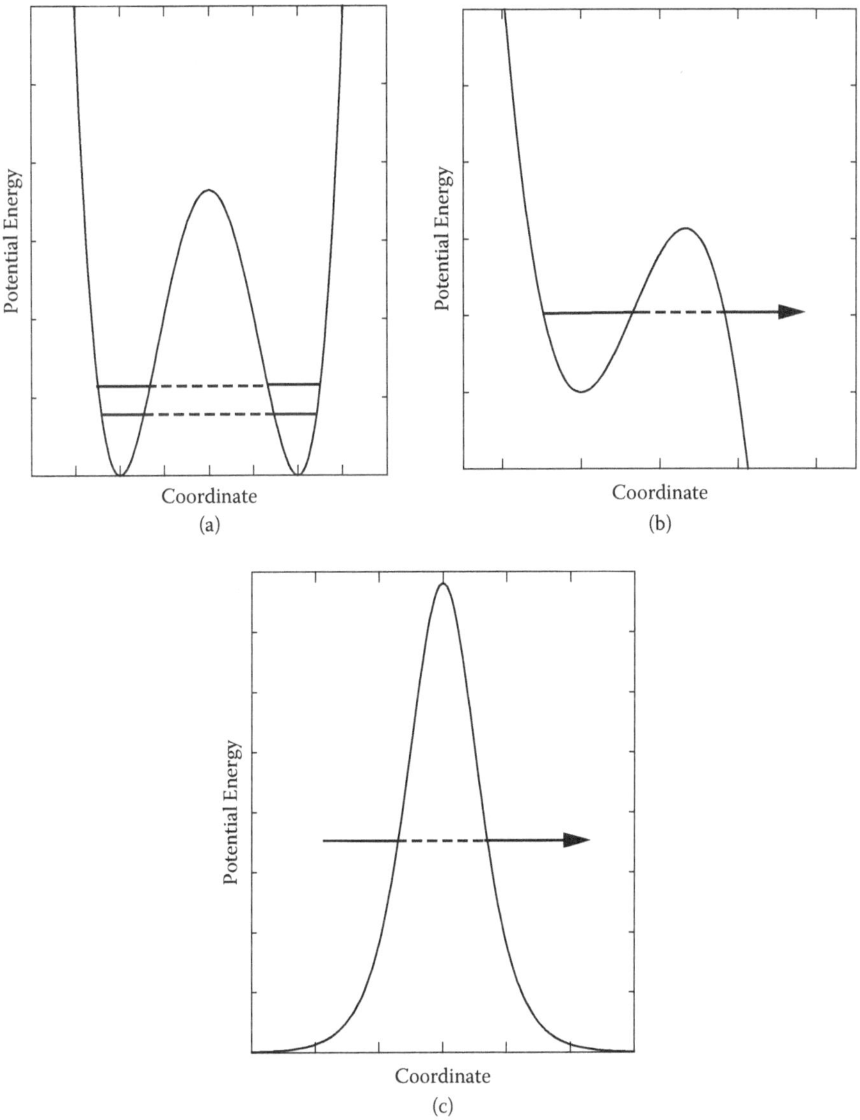

FIGURE 1.1 Three kinds of tunneling phenomena schematically depicted by one-dimensional potential energy curves: (a) tunneling splitting, (b) decay of metastable state, and (c) tunneling in reaction.

the top of that lower adiabatic potential is affected by the coupling with the upper adiabatic potential. Whenever we encounter a potential barrier, we always have to be careful whether any excited state potential is located near the barrier top or why the potential barrier is created.

In addition to the above-mentioned subjects, theories for direct evaluation of reaction rate constants are also discussed for both electronically adiabatic chemical

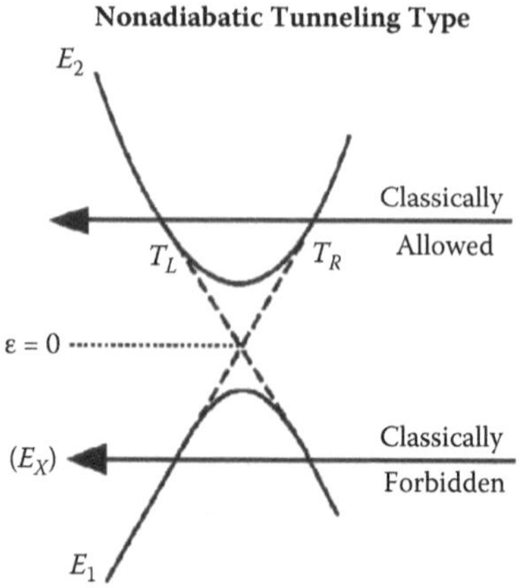

FIGURE 1.2 Nonadiabatic tunneling type of potential curve crossing. Dashed lines represent diabatic states that are coupled to each other. Solid lines are the adiabatic states obtained by diagonalizing the coupled diabataic states.

reactions on a single potential energy surface and nonadiabatic chemical reactions in which two or more adiabatic potential energy surfaces are involved. As is well known, a reaction rate constant is an important physical quantity to characterize chemical reactions. It is desirable to be able to estimate that quantity directly, not from the detailed dynamics (i.e., not from the state-to-state reaction probabilities).

This book is organized as follows. In Chapters 2 and 3 the basic low-dimensional (one- and two-dimensional) theories are reviewed as an introduction to subsequent chapters. In Chapter 4 multidimensionality effects that cannot be interpreted by one-dimensional theory are clarified conceptually. These are endorsed by both numerical demonstrations and experimental observations. An analysis is presented by taking tunneling splittings of the tropolone molecule. The peculiar phenomenon of nonadiabatic tunneling is explained in Chapter 5 and the basic one-dimensional Zhu-Nakamura theory is presented there. Chapter 6 is devoted to the theory of multidimensional tunneling splitting in symmetric double well potential in polyatomic molecules. The above-mentioned WKB-type instanton method (modified WKB method) is presented. A general and practical algorithm to find the instanton trajectory is formulated and the efficient method to include *ab initio* quantum chemical data is explained. Applications to real polyatomic molecules are reported, and the high efficiency and accuracy of the method are demonstrated in Chapter 7. Actual polyatomic molecules treated are hydroperoxy radical (HO_2), vinyl radical (C_2H_3), malonaldehyde ($C_3O_2H_4$), and formic acid dimer [$(DCOOH)_2$]. The algorithm is extended to the decay of metastable states through a potential barrier in Chapter 8. In Chapter 9 tunneling in chemical reactions is discussed. An efficient way of detecting caustics in multidimensional space is presented. Direct evaluation of reaction rate constant is also discussed for both electronically adiabatic and nonadiabatic reactions. Chapter 10 briefly discusses future perspectives and some other topics that are not touched upon in this book.

2 One-Dimensional Theory

2.1 EXACTLY SOLVABLE CASES

The analytical solutions for the following three exactly solvable cases are presented: (1) delta-function barrier, (2) parabolic potential barrier, and (3) Eckart potential barrier. These solutions may be useful for some analyses.

2.1.1 CASE OF DELTA-FUNCTION BARRIER

Let us consider the case of a delta-function barrier $V(x) = \lambda\delta(x)(\lambda > 0)$. The Schrödinger equation to be solved is

$$-\frac{\hbar^2}{2m}\frac{d^2\psi(x)}{dx^2} + \lambda\delta(x)\psi(x) = E\psi(x). \tag{2.1}$$

Since the plane waves $\exp(\pm ikx)$ with $k = \sqrt{2mE}/\hbar$ are the independent solutions at $x \neq 0$, the wave function $\psi(x)$ is expressed as

$$\psi(x) = \begin{cases} -A_r\exp(ikx) + A_l\exp(-ikx), & (x<0), \\ -B_r\exp(ikx) + B_l\exp(-ikx), & (x>0), \end{cases} \tag{2.2}$$

where A_α and $B_\alpha(\alpha = r, l)$ are the coefficients to be determined. The continuity of the wave function at $x = 0$ ($\psi(-0) = \psi(+0)$) requires

$$A_r - A_l = B_r - B_l. \tag{2.3}$$

One more condition is obtained by integrating the Schrödinger equation over the interval $(-\epsilon, \epsilon)$ as

$$-\frac{\hbar^2}{2m}\int_{-\epsilon}^{\epsilon}\psi''(x)dx + \lambda\int_{-\epsilon}^{\epsilon}\delta(x)\psi(x)dx = E\int_{-\epsilon}^{\epsilon}\psi(x)dx, \tag{2.4}$$

where $\epsilon \to 0$ is taken. Since $\int_{-\epsilon}^{\epsilon}\psi''(x)dx = \psi'(+\epsilon) - \psi'(-\epsilon)$, we have

$$A_r + A_l - B_r - B_l = \frac{2m\lambda}{ik\hbar^2}(-A_r + A_l). \tag{2.5}$$

Considering the case that the incident wave comes in from the left side, we have $A_r = 1$ and $B_l = 0$. Putting $A_l = r$ (reflection amplitude) and $B_r = t$ (transmission amplitude) and solving Equations (2.3) and (2.5), we have

$$r = \frac{m\lambda}{i\hbar^2k - m\lambda} \tag{2.6}$$

and

$$t = \frac{i\hbar^2k}{i\hbar^2k - m\lambda}. \tag{2.7}$$

The transfer matrix M defined by

$$\begin{pmatrix} A_r \\ A_l \end{pmatrix} = M \begin{pmatrix} B_r \\ B_l \end{pmatrix} \tag{2.8}$$

is given by

$$M = \begin{pmatrix} 1/t & r^*/t^* \\ r/t & 1/t^* \end{pmatrix}. \tag{2.9}$$

The scattering matrix S defined by

$$\begin{pmatrix} A_l \\ B_r \end{pmatrix} = S \begin{pmatrix} A_r \\ B_l \end{pmatrix} \tag{2.10}$$

is given by

$$S = \begin{pmatrix} r & t \\ t & r \end{pmatrix}. \tag{2.11}$$

The transmission and reflection probabilities are

$$|t|^2 = \frac{\hbar^4 k^2}{\hbar^4 k^2 + m^2 \lambda^2} \tag{2.12}$$

and

$$|r|^2 = \frac{m^2 \lambda^2}{\hbar^4 k^2 + m^2 \lambda^2}. \tag{2.13}$$

2.1.2 Case of Parabolic Potential Barrier

In the case of a parabolic potential barrier, the Schrödinger equation to be solved is given by

$$\left(\frac{d^2}{dx^2} + K^2(x) \right) \psi(x) = 0, \tag{2.14}$$

where

$$K^2(x) = \frac{2m}{\hbar^2}[E - V(x)] \equiv \frac{2m}{\hbar^2} \left(E - V_{\max} + \frac{1}{2}kx^2 \right). \tag{2.15}$$

Introducing the following variable and parameters:

$$z = \left(\frac{4mk}{\hbar^2} \right)^{-1/4} e^{-\pi i/4} x, \tag{2.16}$$

$$\epsilon = \frac{E - V_{\max}}{\hbar \omega^*}, \tag{2.17}$$

and

$$\omega^* = (k/m)^{1/2}, \tag{2.18}$$

the Schrödinger equation is transformed to the Weber equation,

$$\left(\frac{d^2}{dz^2} + i\epsilon - \frac{z^2}{4} \right) \psi(z) = 0. \tag{2.19}$$

The WKB solutions (see Section 2.2) at $x \ll a$ and at $x \gg b$ are expressed, respectively, as

$$\psi(x) \overset{x\ll a}{\simeq} [k(x)]^{-1/2}\left(-C_a^{'}\exp\left[i\int_a^x k(x)dx\right] + C_a^{''}\exp\left[-i\int_a^x k(x)dx\right]\right) \tag{2.20}$$

and

$$\psi(x) \overset{x\gg b}{\simeq} [k(x)]^{-1/2}\left(-C_b^{'}\exp\left[i\int_b^x k(x)dx\right] + C_b^{''}\exp\left[-i\int_b^x k(x)dx\right]\right). \tag{2.21}$$

It should be noted that the coefficients of the wave running to the right are taken to be $-C_a'$ and $-C_b'$ as in Equation (2.2). By using the asymptotic expressions of Weber functions and the above WKB solutions, the connection between the coefficients $(C_a^{'}, C_a^{''})$ and $(C_b^{'}, C_b^{''})$ can be obtained as

$$\begin{pmatrix} C_a^{'} \\ C_a^{''} \end{pmatrix} = \begin{pmatrix} \frac{\sqrt{2\pi}}{\Gamma(1/2-i\epsilon)}\exp[-\frac{\pi}{2}\epsilon + i\epsilon - i\epsilon\ln|\epsilon|] & -i\exp[-\pi\epsilon] \\ i\exp[-\pi\epsilon] & \frac{\sqrt{2\pi}}{\Gamma(1/2+i\epsilon)}\exp[-\frac{\pi}{2}\epsilon - i\epsilon + i\epsilon\ln|\epsilon|] \end{pmatrix} \begin{pmatrix} C_b^{'} \\ C_b^{''} \end{pmatrix} \equiv M\begin{pmatrix} C_b^{'} \\ C_b^{''} \end{pmatrix}, \tag{2.22}$$

where $\Gamma(z)$ is the gamma function. Using the relation

$$\Gamma(1/2+i\epsilon) = \sqrt{2\pi}\exp\left[-\frac{\pi}{2}\epsilon\right](1+\exp[-2\pi\epsilon])^{-1/2}\exp[i\arg\Gamma(1/2+i\epsilon)], \tag{2.23}$$

the final expression of the transfer matrix M can be obtained as

$$M = \begin{pmatrix} (1+\exp[-2\pi\epsilon])^{1/2}\exp[i\phi(\epsilon)] & -i\exp[-\pi\epsilon] \\ i\exp[-\pi\epsilon] & (1+\exp[-2\pi\epsilon])^{1/2}\exp[-i\phi(\epsilon)] \end{pmatrix}, \tag{2.24}$$

where the phase $\phi(\epsilon)$ is defined as

$$\phi(\epsilon) = \arg\Gamma(1/2+i\epsilon) + \epsilon - \epsilon\ln|\epsilon|. \tag{2.25}$$

The coefficients $C_a^{'}$ and $C_b^{''}$ represent the incoming waves to the potential barrier and the coefficients $C_b^{'}$ and $C_a^{''}$ represent the outgoing waves from the barrier. The matrix S^T that connects these sets of coefficients has physical meaning of scattering matrix defined as

$$\begin{pmatrix} C_b^{'} \\ C_a^{''} \end{pmatrix} = S^T\begin{pmatrix} C_a^{'} \\ C_b^{''} \end{pmatrix}. \tag{2.26}$$

This tunneling scattering matrix can be obtained easily from the above expression of M-matrix as

$$S^T = \exp[-i\phi(\epsilon)](1+\exp[-2\pi\epsilon])^{-1/2}\begin{pmatrix} 1 & i\exp[-\pi\epsilon] \\ i\exp[-\pi\epsilon] & 1 \end{pmatrix}. \tag{2.27}$$

The diagonal (off-diagonal) elements of this matrix represent tunneling (reflection) amplitudes. At $E \leq V_{max}$ the parameter ϵ is negative [$\epsilon = -|\epsilon|$] [see Equation (2.17)] and the tunneling probability is given by

$$P_{tunnel} = \frac{\exp[-2\pi|\epsilon|]}{1 + \exp[-2\pi|\epsilon|]}. \tag{2.28}$$

In the deep tunneling limit $E \ll V_{max}$, i.e., $|\epsilon| \gg 1$, this becomes the same as the Gamov formula. On the other hand, at $E \geq V_{max}$, $\epsilon(= |\epsilon|)$ is positive [see Equation (2.17)] and the transmission probability becomes

$$P_{tunnel} = \frac{\exp[2\pi|\epsilon|]}{1 + \exp[2\pi|\epsilon|]}. \tag{2.29}$$

In the high-energy limit $E \gg V_{max}$, the probability naturally goes to unity.

2.1.3 Case of Eckart Potential Barrier

The Schrödinger equation to be solved is given by [49]

$$\left(\frac{d^2}{dx^2} + K^2(x)\right)\psi(x) = 0, \tag{2.30}$$

where

$$K^2(x) = \frac{2m}{\hbar^2}[E - V(x)] \equiv \frac{2m}{\hbar^2}\left[E - U_0\text{sechyp}^2(\alpha x)\right]. \tag{2.31}$$

The solution of this equation is given in the textbook of Landau and Lifshitz [50]. With the variable changes

$$u = \frac{1}{2}(1 - \xi) = \frac{1}{2}[1 - \tanh(\alpha x)] \tag{2.32}$$

and

$$\psi(x) = \left[1 - (1 - 2u)^2\right]^{-i\epsilon/2} w, \tag{2.33}$$

we obtain the differential equation

$$u(1 - u)\frac{d^2w}{du^2} + (1 - i\epsilon)(1 - 2u)\frac{dw}{du} + [i\epsilon(1 - i\epsilon) + s(s + 1)]w = 0, \tag{2.34}$$

where

$$\epsilon = \sqrt{\frac{2mE}{\alpha^2\hbar^2}} = \frac{k}{\alpha} \tag{2.35}$$

and

$$s = \begin{cases} \frac{1}{2}\left[-1 + \sqrt{1 - 8mU_0/(\alpha^2\hbar^2)}\right] \equiv -\frac{1}{2} + s_1 & \text{for } 8mU_0/(\alpha^2\hbar^2) \leq 1, \\ \frac{1}{2}\left[-1 + i\sqrt{8mU_0/(\alpha^2\hbar^2) - 1}\right] \equiv -\frac{1}{2} + is_0 & \text{for } 8mU_0/(\alpha^2\hbar^2) \geq 1. \end{cases} \tag{2.36}$$

The physically meaningful solution of Equation (2.34) is given by the hypergeometric function as

$$w = F\big(-i\epsilon - s, -i\epsilon + s + 1, 1 - i\epsilon : [1 - \tanh(\alpha x)]/2\big). \tag{2.37}$$

For $x \to +\infty$, we have $\xi \to 1, u \to 0, w \to 1$, and

$$\psi(x) \to [1 - (1 - 2u)^2]^{-i\epsilon/2} \to 2^{-i\epsilon} e^{ikx}, \tag{2.38}$$

which represents the tunneling wave moving to the right. For $x \to -\infty$, on the other hand, we have $\xi \to -1, u \to 1$ and thus

$$\psi(x) \to 2^{-i\epsilon}[Ae^{ikx} + Be^{-ikx}], \tag{2.39}$$

where

$$A = \frac{\Gamma(1 - i\epsilon)\Gamma(-i\epsilon)}{\Gamma(-i\epsilon - s)\Gamma(-i\epsilon + s + 1)} \equiv |A| \exp(i\theta_A) \tag{2.40}$$

and

$$B = \frac{\Gamma(1 - i\epsilon)\Gamma(i\epsilon)}{\Gamma(1 + s)\Gamma(-s)} \equiv |B| \exp(i\theta_B). \tag{2.41}$$

In order to derive Equation (2.39), the following relation of the hypergeometric functions is used:

$$\begin{aligned} F(\alpha, \beta, \gamma : z) = {} & \frac{\Gamma(\gamma)\Gamma(\gamma - \alpha - \beta)}{\Gamma(\gamma - \alpha)\Gamma(\gamma - \beta)} F(\alpha, \beta, \alpha + \beta + 1 - \gamma : 1 - z) \\ & + \frac{\Gamma(\gamma)\Gamma(\alpha + \beta - \gamma)}{\Gamma(\alpha)\Gamma(\beta)} (1 - z)^{\gamma - \alpha - \beta} \\ & \times F(\gamma - \alpha, \gamma - \beta, \gamma + 1 - \alpha - \beta : 1 - z). \end{aligned} \tag{2.42}$$

The transmission amplitude T and the reflection amplitude R are obtained as follows:

$$T = \frac{1}{A} = |T| \exp(-i\theta_A) \tag{2.43}$$

and

$$R = \frac{B}{A} = |R| \exp(i\theta_B - i\theta_A). \tag{2.44}$$

The tunneling probability is given by

$$P = |T|^2 = \begin{cases} \frac{\sinh^2(\pi\epsilon)}{\sinh^2(\pi\epsilon) + \cos^2(\pi s_1)} & \text{for } s = -1/2 + s_1, \\ \frac{\sinh^2(\pi\epsilon)}{\sinh^2(\pi\epsilon) + \cosh^2(\pi s_0)} & \text{for } s = -1/2 + i s_0. \end{cases} \tag{2.45}$$

2.2 WKB APPROXIMATION AND CONNECTION FORMULA

The WKB approximation, which is sometimes called the JWKB approximation, paying respect to the contribution of mathematician H. Jeffreys, provides us with the analytical expression of wave function for a particle moving in a general potential $V(x)$ as

$$\psi(x) \simeq \frac{1}{\sqrt{k(x)}} \exp[\pm \frac{i}{\hbar} \int_{x_0}^{x} p(x)dx], \tag{2.46}$$

where x_0 is the turning point $[E = V(x_0)]$ and

$$k(x) = \frac{p(x)}{\hbar} = \frac{1}{\hbar}\sqrt{2m[E - V(x)]}. \tag{2.47}$$

Since this wave function becomes the quantum mechanically exact solution—namely, plane wave, when the potential $V(x)$ does not depend on the coordinate x and is constant—the WKB approximation holds well when the potential weakly depends on the coordinate. That is to say, the condition that the above WKB approximation holds well is given by

$$\frac{1}{2\pi}\left|\frac{d\lambda(x)}{dx}\right| \ll 1, \tag{2.48}$$

where $\lambda(x)$ is the coordinate-dependent de Broglie wavelength. This approximation is useful, since it gives an analytical expression of wave function for any potential $V(x)$ as far as the above condition is satisfied. At the turning point x_0, however, it breaks down, since $k(x) \propto \sqrt{x - x_0}$ there. The turning point is nothing but the boundary between the classically allowed and forbidden regions. In the classically forbidden region, the two independent WKB solutions are exponentially decaying and growing functions, as can be easily guessed from Equation (2.46). In order to obtain a physical solution that satisfies physical boundary conditions, the WKB functions that hold in each region should be connected. This connection is one of the most important problems of the WKB approximation. As an example, let us consider the one-dimensional tunneling problem depicted in Figure 1.1(c). Suppose the incident wave comes in from the left and the turning points on the left and right sides are denoted as $x = a$ and $x = b$, respectively.

At the turning point $x = b$, the following two connection formulas hold:

$$\frac{c}{2\sqrt{|k(x)|}} \exp\left[-\left|\int_b^x k(x)dx\right|\right] \leftrightarrow \frac{c}{\sqrt{k(x)}} \sin\left[\left|\int_b^x k(x)dx + \pi/4\right|\right] \tag{2.49}$$

and

$$\frac{-ic}{\sqrt{|k(x)|}} \exp\left[+\left|\int_b^x k(x)dx\right|\right] \leftrightarrow \frac{c}{\sqrt{k(x)}} \exp\left[i\int_b^x k(x)dx - \pi i/4\right], \tag{2.50}$$

where c is a constant. Equation (2.49) tells that the stationary wave in the classically allowed region is connected to the exponentially decaying function with the coefficient $1/2$. On the other hand, Equation (2.50) means that the outgoing wave in the classically allowed region is connected to the exponentially growing wave in the classically forbidden region. The factor $-i$ in the classically forbidden region and the phase $-\pi i/4$ in the classically allowed region should be noted. Suppose the incident wave comes in from the left at $x \ll a$. This incident wave is expressed as

$$\frac{1}{2i\sqrt{p}}\left(\exp\left[i\int_x^a k(x)dx + \pi i/4\right] - \exp\left[-i\int_x^a k(x)dx - \pi i/4\right]\right)$$
$$= \frac{1}{p}\sin\left(\int_x^a k(x)dx + \pi/4\right), \tag{2.51}$$

where the second (first) exponential function represents the incident (reflected) wave. Here the tunneling probability is assumed to be small and the amplitude of the reflected wave is set to be unity. Now we apply the connection formula Equation (2.49) and we have the following wave inside the potential barrier at $a < x < b$:

$$\frac{1}{2\sqrt{|p(x)|}}\exp\left[-\left|\int_a^x k(x)dx\right|\right]. \tag{2.52}$$

This can be rewritten as

$$\exp\left[-\left|\int_a^b k(x)dx\right|\right] \times \frac{1}{2\sqrt{|p(x)|}}\exp\left[\left|\int_b^x k(x)dx\right|\right]. \tag{2.53}$$

Here we can apply the connection formula Equation (2.50) to the last exponential function in the above expression and we have the wave function in the classically allowed region at $x > b$ as

$$\exp\left[-\left|\int_a^b k(x)dx\right|\right] \times \frac{i}{2\sqrt{p(x)}}\exp\left[i\int_b^x k(x)dx - \pi i/4\right]. \tag{2.54}$$

The square of the first term of this expression gives the tunneling probability as

$$P_{\text{tunnel}} \simeq \exp\left[-2\left|\int_a^b k(x)dx\right|\right]. \tag{2.55}$$

This is the well-known Gamov formula of tunneling probability. It is now well understood that the connection formulas of Equations (2.49) and (2.50) are crucial. These formulas can be obtained from the Airy function, since the potential in the vicinity of the turning point can be approximated by a linear function of x for which the Airy function gives the exact analytical solution.

2.3 COMPARISON EQUATION METHOD

The Gamov formula derived above is a deep tunneling approximation at $E \ll V_{\text{max}}$, where V_{max} is the potential barrier top. Actually, Equation (2.55) gives 1.0 at $E = V_{\text{max}}$, but the correct transmission probability at $E \simeq V_{\text{max}}$ is $\sim 1/2$. This error is obvious, since the linear potential approximation does not hold in the vicinity of the potential barrier top. It is better to use a quadratic potential approximation. Instead of directly using the parabolic potential approximation, a more general comparison equation method is explained here [51]. Suppose we know the exact solution of the following differential equation:

$$\left[\frac{d^2}{dx^2} + J^2(\xi)\right]\phi(\xi) = 0. \tag{2.56}$$

We try to solve a more general equation that has the same analytical structure,

$$\left[\frac{d^2}{dx^2} + K^2(x)\right]\psi(x) = 0, \tag{2.57}$$

in the form

$$\psi_{app}(x) = A(x)\phi[\xi(x)]. \tag{2.58}$$

In order for the function $\psi_{app}(x)$ to be a good approximation to $\psi(x)$, the WKB solutions of Equations (2.56) and (2.57),

$$\psi_{WKB}(x) = K^{-1/2}(x)\exp\left[i\int_{x_0}^{x} K(x')dx'\right] \tag{2.59}$$

and

$$\phi_{WKB}(\xi) = J^{-1/2}(\xi)\exp\left[i\int_{\xi_0}^{\xi} J(\xi')d\xi'\right], \tag{2.60}$$

should coincide. In other words, the following relations should be satisfied:

$$\int_{x_0}^{x} K(x')dx' = \int_{\xi_0}^{\xi} J(\xi')d\xi' \tag{2.61}$$

and

$$A(x) = \left(\frac{J(\xi)}{K(x)}\right)^{1/2} = \left(\frac{d\xi}{dx}\right)^{-1/2}. \tag{2.62}$$

The approximate function $\psi_{app}(x)$ can be shown to satisfy the equation

$$\left[\frac{d^2}{dx^2} + K^2(x) + \gamma(x)\right]\psi_{app}(x) = 0, \tag{2.63}$$

where

$$\gamma(x) = -\left(\frac{d\xi}{dx}\right)^{1/2}\frac{d^2}{dx^2}\left(\frac{d\xi}{dx}\right)^{-1/2} = -\left(\frac{K(x)}{J(\xi)}\right)^{1/2}\frac{d^2}{dx^2}\left(\frac{K(x)}{J(\xi)}\right)^{-1/2}. \tag{2.64}$$

Thus, $\psi_{app}(x)$ can be a good approximation, when the following condition is satisfied:

$$\gamma(x) \ll K^2(x). \tag{2.65}$$

The important question about this comparison equation method is what "the same analytical structure" means. In order for the function $\gamma(x)$ given by Equation (2.64) not to diverge, the order and the number of zeros of $K(x)$ should be the same as those of $J(\xi)$. That is to say, if the conditions

$$K(x_j) = J(\xi_j) = 0 \quad (j = 1, 2, 3, , ,) \tag{2.66}$$

and

$$\int_{x_j}^{x_{j+1}} K(x)dx = \int_{\xi_j}^{\xi_{j+1}} J(\xi)d\xi \tag{2.67}$$

are satisfied, then we can construct a good approximate solution $\psi_{app}(x)$ from the solution $\phi(\xi)$ of the standard Equation (2.56).

In the case of a general potential that has "the same analytical structure" as that of the parabolic potential barrier, we can use the comparison equation method explained above. Equation (2.67) requires

$$\int_a^b K(x)dx = \int_{-2(i\epsilon)^{1/2}}^{2(i\epsilon)^{1/2}} (-z^2/4 + i\epsilon)^{1/2} dz = -i\pi\epsilon. \tag{2.68}$$

Thus we obtain

$$\epsilon = \begin{cases} -\frac{1}{\pi}\int_a^b |K(x)|dx & \text{for } E \leq V_{\max}, \\ -\frac{1}{\pi i}\int_{x_1}^{x_2} K(x)dx & \text{for } E \geq V_{\max}, \end{cases} \tag{2.69}$$

where x_1 and x_2 are the complex turning points that satisfy $K(x_j) = 0$ $(j = 1, 2)$ at $E \geq V_{\max}$. With this replacement of ϵ we can use the results obtained in Section 2.1.2. The matrices M of Equation (2.24) and S of Equation (2.27) can be used together with ϵ given by Equation (2.69).

Unfortunately, this method cannot be used in the case of an Eckart potential barrier, discussed in Section 2.1.3. The corresponding appropriate comparison equation method has not yet been developed in this case.

2.4 DIAGRAMMATIC TECHNIQUE

Many chemical dynamic processes can be decomposed into basic events and described as a sequence of those events. The basic events are (1) wave propagation along adiabatic potential without any transition, (2) wave reflection at turning point, (3) potential barrier transmission and reflection, (4) nonadiabatic transition at avoided crossing point, and (5) nonadiabatic tunneling and reflection at the energy lower than the bottom of the upper adiabatic potential. These are expressed by the diagrams depicted in Figures 2.1–2.5.

Various one-dimensional processes can be expressed by connecting these diagrams and can be described by combining the appropriate semiclassical matrices. This technique is called diagrammatic technique [48,52]. When we write the semiclassical wave function on the adiabatic potential $E_n(x)$ with a as a reference point as

$$\psi_n(x) \simeq -V_n'\phi_n^{(+)}(x:a) + V_n''\phi_n^{(-)}(x:a) \tag{2.70}$$

with

$$\phi_n^{(\pm)}(x:a) = \sqrt{\frac{m}{2\pi k_n(x)}} \exp\left(\pm i \int_a^x k_n(x)dx\right), \tag{2.71}$$

the connection of the coefficients is provided as follows for each basic element.

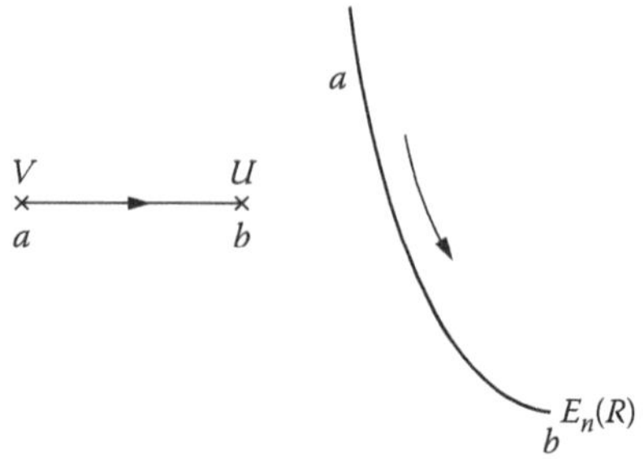

FIGURE 2.1 Diagram for wave propagation from a to b along the potential $E_n(R)$.

1. **Wave propagation (see Figure 2.1)**

$$\mathcal{U} = \begin{pmatrix} \exp[i\gamma_n(a,b)] & 0 \\ 0 & \exp[-i\gamma_n(a,b)] \end{pmatrix} \mathcal{V} \equiv P_{ab}\mathcal{V} \ (a < b), \quad (2.72)$$

where

$$\gamma_n(a,b) = \int_a^b k_n(x)dx, \quad (2.73)$$

$\mathcal{V}$ is a column vector with components V' and V'', and $\mathcal{U}$ is the similar vector that represents the wave function with b as a reference point.

2. **Wave reflection at turning point (see Figure 2.2)**

$$\mathcal{V} = \begin{pmatrix} e^{\pi i/4} & 0 \\ 0 & e^{-\pi i/4} \end{pmatrix} \begin{pmatrix} A \\ A \end{pmatrix}, \quad (2.74)$$

where A simply represents the wave amplitude. If the turning point is located on the right side, the signs of $\pi i/4$ should be interchanged. If we eliminate the amplitude A, then we have $V' = \exp(i\pi/2)V''$.

3. **Potential barrier transmission and reflection (see Figure 2.3)**

$$\mathcal{V} = \begin{pmatrix} \sqrt{1+\exp[-2\pi\epsilon]}e^{i\phi} & -i\exp[-\pi\epsilon] \\ i\exp[-\pi\epsilon] & \sqrt{1+\exp[-2\pi\epsilon]}e^{-i\phi} \end{pmatrix} \mathcal{U} \equiv M\mathcal{U}. \quad (2.75)$$

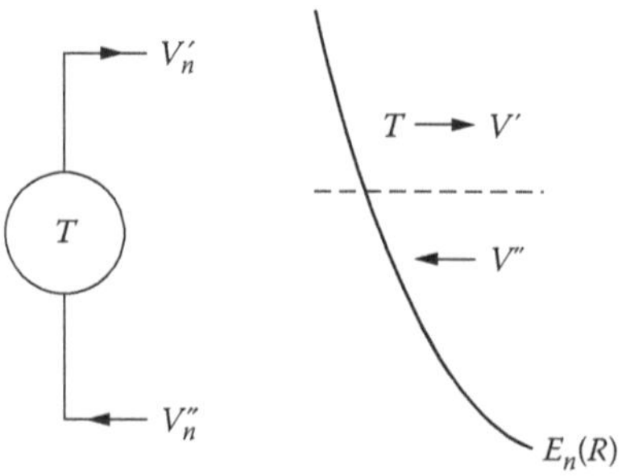

FIGURE 2.2 Diagram for wave reflection at the turning point T.

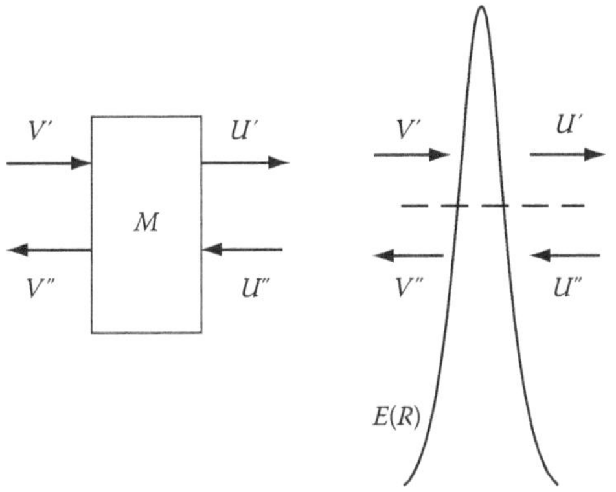

FIGURE 2.3 Diagram for potential barrier transmission and reflection.

Here we have used the comparison equation method for the case of quadratic potential barrier. This transfer matrix M is the same as Equation (2.24), and ϕ and ϵ are given by Equations (2.25) and (2.69).

4. **Nonadiabatic transition at avoided crossing point (see Figure 2.4)**

$$\mathcal{U}'' = I\mathcal{V}'' \text{ and } \mathcal{V}' = O\mathcal{U}', \tag{2.76}$$

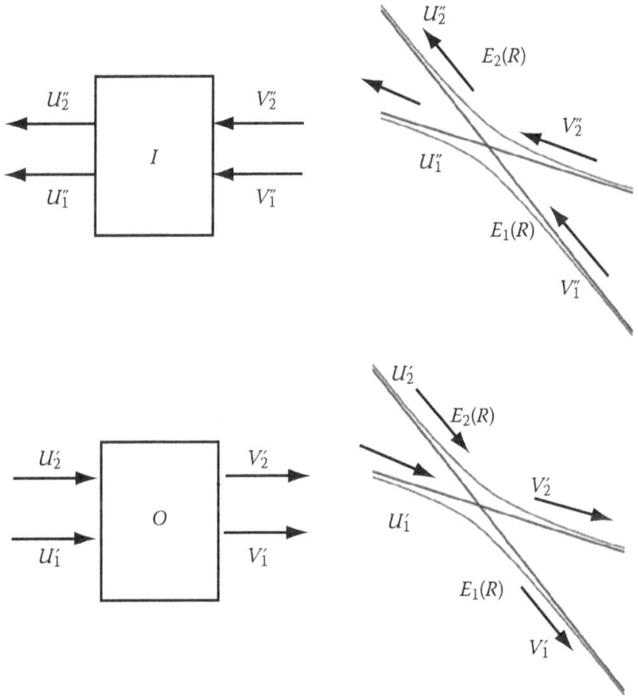

FIGURE 2.4 Diagram for nonadiabatic transition. Matrix O is a transpose of I ($O = I'$).

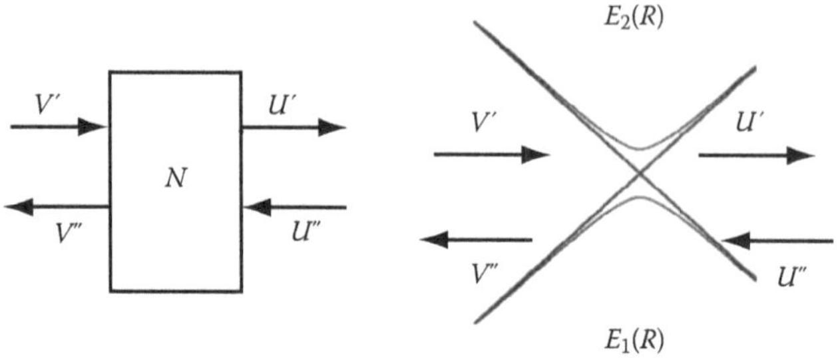

FIGURE 2.5 Diagram for nonadiabatic tunneling.

where $\mathcal{U}'$ is a column vector with components U_1' and U_2', and the other vectors have similar meanings. When the energy is lower than the crossing point E_X, we have

$$\mathcal{V}' = S^{(R)}\mathcal{V}'', \tag{2.77}$$

where $S^{(R)}$ is the so-called reduced scattering matrix, which is a scattering matrix with elastic scattering phase parts removed from the total scattering matrix.

5. **Nonadiabatic tunneling at** $E \leq E_b$ (bottom of the upper adiabatic potential) **(see Figure 2.5)**

$$\mathcal{U} = N\mathcal{V}, \tag{2.78}$$

where N is a transfer matrix.
Since the nonadiabatic transition is not a main subject in this book, the expressions of the matrices I, $O(=$ transpose of $I)$, $S^{(R)}$, and N are not given here. The reader interested in these should refer to Reference [48], and especially Appendix A of this book. The nonadiabatic tunneling problem is, however, discussed in Chapter 5.

2.5 INSTANTON THEORY AND MODIFIED WKB METHOD

In the first part of this section, the instanton theory [2] is explained by taking the motion of a particle of mass m in one-dimensional potential $V(x)$. Tunneling splitting in a symmetric double well potential and decay of metastable state by tunneling through a potential barrier are employed as examples. In the second subsection, it is shown that the results can be reproduced by the WKB method with slight modification.

2.5.1 Instanton Theory

The instanton theory was invented in the field theory by introducing imaginary time to Minkowski space-time. The instanton is the classical object in the Euclidean space-time that gives a finite action. The instanton is also called pseudo-particle. Here the theory is explained by following Coleman [2]. See also References [17,39,43,46].

Let us consider the transition amplitude from position x_i to x_f under the Hamiltonian H,

$$\langle x_f|e^{-iHt/\hbar}|x_i\rangle = N\int D[x]e^{iS_c/\hbar}, \tag{2.79}$$

where S_c is the action, N is a normalization factor, and $D[x]$ means Feynman's path integral over all functions $x(t)$, which satisfies the boundary conditions $x(-t/2) = x_i$ and $x(t/2) = x_f$. For simplicity of formulation, we assume for a moment that the particle mass is equal to unity. The mass m is recovered in the final expression. Here we introduce the imaginary time $t = -iT$; then the above expression becomes

$$\langle x_f|e^{-HT/\hbar}|x_i\rangle = N\int D[x]e^{-S/\hbar}, \tag{2.80}$$

where S is the Euclidean action defined by

$$S = \int_{-T/2}^{T/2} d\tau\left[\frac{1}{2}\left(\frac{dx}{d\tau}\right)^2 + V\right]. \tag{2.81}$$

The reason we consider the amplitude given by Equation (2.80) is as follows. If we expand this amplitude by using a complete set of energy eigenstates,

$$H|n\rangle = E_n|n\rangle, \tag{2.82}$$

then

$$\langle x_f|e^{-HT\hbar}|x_i\rangle = \sum_n e^{-E_nT/\hbar}\langle x_f|n\rangle\langle n|x_i\rangle. \tag{2.83}$$

At $T \to \infty$, the leading term gives the lowest energy and the corresponding wave function.

Now we try to evaluate the right side of Equation (2.80) by using the expansion

$$x(\tau) = \bar{x}(\tau) + \sum_n c_n x_n(\tau), \tag{2.84}$$

where $\bar{x}(\tau)$ is a certain function satisfying the boundary condition and $\{x_n(\tau)\}$ is a complete set of orthonormal functions vanishing at the boundaries,

$$\int_{-T/2}^{T/2} d\tau x_n(\tau)x_m(\tau) = \delta_{nm}, \quad x_n(\pm T/2) = 0. \tag{2.85}$$

Then, the integration measure $D[x]$ is given by

$$D[x] = \prod_n (2\pi\hbar)^{-1/2} dc_n. \tag{2.86}$$

Here we use the semiclassical approximation (small $\hbar$); namely, we use the stationary phase approximation, assuming that there is only one stationary point, which is denoted by $\bar{x}$. The variation of the action S for the change of $x \to \bar{x} + \delta x$ is given by using the partial integration method as

$$S(\bar{x}+\delta x) = S(\bar{x}) + \int d\tau\left[-\frac{d^2\bar{x}}{d\tau^2} + V'(\bar{x})\right]\delta x + \int d\tau\delta x\left[-\frac{d^2\delta x}{d\tau^2} + V''\delta x\right] + \cdots. \tag{2.87}$$

The stationary solution $\bar{x}$, which is called instanton trajectory, satisfies the equation

$$\frac{\delta S}{\delta \bar{x}} = -\frac{d^2\bar{x}}{d\tau^2} + V'(\bar{x}) = 0. \tag{2.88}$$

It should be noted that this is a classical equation of motion in the *inverted* potential $-V(x)$. Let us choose the functions $\{x_n(\tau)\}$ to be eigenfunctions of the second variational derivative of S at $\bar{x}$ [see Equation (2.87)],

$$-\frac{d^2x_n(t)}{d\tau^2} + V''(\bar{x})x_n(\tau) = \lambda_n x_n(\tau). \tag{2.89}$$

Since

$$S = S(\bar{x}) + \frac{1}{2}\sum_n \lambda_n c_n^2 \equiv S_0 + \frac{1}{2}\sum_n \lambda_n c_n^2, \tag{2.90}$$

the integral in Equation (2.80) can be carried out by Gaussian integrals and we have

$$\begin{aligned}\langle x_f|e^{-HT/\hbar}|x_i\rangle &= Ne^{-S_0/\hbar}\prod_n \lambda_n^{-1/2}\\ &= Ne^{-S_0/\hbar}\left[\det\left(-d^2/d\tau^2 + V''(\bar{x})\right)\right]^{-1/2},\end{aligned} \tag{2.91}$$

where the second equation gives the definition of the determinant of differential operator; namely, det[...] means the infinite product of the eigenvalues of the differential operator in the bracket. It is assumed here that all eigenvalues are positive. This point is discussed later.

Now, let us consider the double well potential $V(x)$ (see Figure 2.9), which is symmetric $V(-x) = V(x)$ and vanishes at minima with frequency ω there, namely, $V(x = \pm x_0) = 0$ and $V''(x = \pm x_0) = \omega^2$. First we have to find solutions that satisfy Equation (2.88). There are two obvious solutions in which the particle stays forever at one of the tops of the two hills. But there is another important solution that starts from the top of one hill, say at $x = -x_0$ at time $\tau = -T/2$, and moves to the top of the other hill at $x = x_0$ at time $\tau = T/2$. From Equation (2.88) we obtain

$$\frac{d\bar{x}}{d\tau} = \sqrt{2V} \tag{2.92}$$

or

$$\tau = \tau_1 + \int_0^x dx'(2V)^{-1/2}, \tag{2.93}$$

where τ_1 is an integration constant time at which x vanishes. This solution is called "an instanton with center at τ_1." Solutions that go from x_0 to $-x_0$ are called "anti-instantons." When x approaches x_0 at large τ, Equation (2.92) can be approximated as $dx/d\tau = \omega(x_0 - x)$, which gives $(x_0 - x) \propto e^{-\omega\tau}$ at large τ. This indicates that instantons are well-localized objects. From Equations (2.81) and (2.92), the action S_0 is given by

$$S_0 = \int d\tau\left[\frac{1}{2}(dx/d\tau)^2 + V\right] = \int d\tau(dx/d\tau)^2 = \int_{-x_0}^{x_0} dx\sqrt{2V(x)}. \tag{2.94}$$

Since the instantons are well localized, strings of widely separated instantons and anti-instantons can also be solutions and we have to sum up all of them. Suppose there are well-separated n objects (instantons or anti-instantons), the action is nS_0, and the contribution to the transition amplitude from one object is the same as that for a single harmonic potential and is given by (see Appendix A),

$$N\left[\det\left(-\frac{d^2}{d\tau^2}+\omega^2\right)\right]^{-1/2}=\left(\frac{\omega}{\pi\hbar}\right)^{1/2}e^{-\omega T}. \tag{2.95}$$

The interval between the two objects is actually not infinity and thus the correction to this is introduced as

$$\left(\frac{\omega}{\pi\hbar}\right)^{1/2}e^{-\omega T}[B/2]^n, \tag{2.96}$$

where B is derived below by demanding that this formula gives the right answer for one instanton. The functional integration should be carried out over the locations of the centers $\{\tau_n\}$ as

$$\int_{-T/2}^{T/2}d\tau_1\int_{-T/2}^{\tau_1}d\tau_2\cdots\int_{-T/2}^{\tau_{n-1}}d\tau_n=T^n/n!. \tag{2.97}$$

If we start at $-x_0(x_0)$ and come back to $x_0(-x_0)$, then n must be an odd number and we can obtain the corresponding transition amplitudes as

$$\begin{aligned}\langle x_0|e^{-HT/\hbar}|-x_0\rangle &= \langle -x_0|e^{-HT/\hbar}|x_0\rangle\\ &= \left(\frac{\omega}{\pi\hbar}\right)^{1/2}e^{-\omega T}\sum_{\text{odd } n}\frac{(BTe^{-S_0/\hbar}/2)^n}{n!}.\end{aligned} \tag{2.98}$$

The amplitudes $\langle -x_0|e^{-HT/\hbar}|-x_0\rangle=\langle x_0|e^{-HT/\hbar}|x_0\rangle$ are given by the same expression with the summation over even n. Denoting the splitted ground states as $|0\rangle$ and $|1\rangle$ with energies E_0 and $E_1(E_0<E_1)$, then we have

$$\begin{aligned}|\langle 1|\pm x_0\rangle|^2 &= |\langle 0|\pm x_0\rangle|^2\\ &= \langle x_0|0\rangle\langle 0|-x_0\rangle=-\langle x_0|1\rangle\langle 1|-x_0\rangle=\frac{1}{2}\left(\frac{\omega}{\pi\hbar}\right)^{1/2}.\end{aligned} \tag{2.99}$$

Since at $T\to\infty$

$$\begin{aligned}\exp[-E_0(E_1)T/\hbar]=\sqrt{\pi\hbar/\omega}\ \Big[&\langle -x_0|e^{-HT/\hbar}|-x_0\rangle\\ &+(-)\langle x_0|e^{-HT/\hbar}|-x_0\rangle\Big],\end{aligned} \tag{2.100}$$

we have finally

$$E_0(E_1)=\frac{\hbar\omega}{2}-(+)\hbar\frac{B}{2}e^{-S_0/\hbar}. \tag{2.101}$$

Thus the tunneling splitting Δ_0 of the ground state is

$$\Delta_0\equiv E_1-E_0=\hbar Be^{-S_0/\hbar}. \tag{2.102}$$

We now turn to the evaluation of B. This is related to the zero eigenvalue solution as follows. Because of the time translation invariance (i.e., the invariance with respect to the time shift) the eigenvalue Equation (2.89) has a solution with zero eigenvalue ($\lambda_1 = 0$) as

$$x_1 = S_0^{-1/2}\frac{d\bar{x}}{d\tau}. \tag{2.103}$$

This can be easily proved by using Equation (2.88) satisfied by $\bar{x}(t)$. The normalization factor is obtained from the normalization of $x_1(\tau)(\int x_1^2 d\tau = 1)$ and Equation (2.94). The integration over the corresponding coefficient c_1 [see Equation (2.84)] diverges. This integration is the same as the integration over τ_1 in Equation (2.97) apart from a proportional coefficient. From Equation (2.84) we obtain

$$dx = x_1 dc_1 \tag{2.104}$$

and the change dx induced by a change of the center location τ_1 is equal to

$$dx = (d\bar{x}/d\tau)d\tau_1. \tag{2.105}$$

Hence, we have [see Equation (2.103)]

$$dc_1 = S_0^{1/2} d\tau_1 \tag{2.106}$$

and thus the one-instanton contribution to the transition amplitude is given by

$$\langle x_0|e^{-HT/\hbar}|-x_0\rangle = NT(S_0/2\pi\hbar)^{1/2}e^{-S_0/\hbar}\left[\det'\left(-\frac{d^2}{d\tau^2}+V''(\bar{x})\right)\right]^{-1/2}, \tag{2.107}$$

where det′ indicates that the zero eigenvalue is omitted when computing the determinant. Comparing this to the one-instanton expression, namely, the first ($n = 1$) term in Equation (2.98), we have

$$B = \sqrt{\frac{2S_0}{\pi\hbar}}\left[\frac{\det(-\frac{d^2}{d\tau^2}+\omega^2)}{\det'(-\frac{d^2}{d\tau^2}+V''(\bar{x}))}\right]^{1/2}. \tag{2.108}$$

Now we further transform the expression of Equation (2.107) [43]. Let us consider the quantity

$$\frac{\det[-\frac{d^2}{d\tau^2}+\omega^2]_{T/2}\lambda(T/2)}{\det[-\frac{d^2}{d\tau^2}+V''(\bar{x})]_{T/2}}, \tag{2.109}$$

where $\lambda(T/2)$ is the lowest eigenvalue of the operator in the denominator so that the denominator becomes the primed determinant at $T \to \infty$ as Equation (2.108) requires. As is proved in Reference [2], Equation (2.109) (see Appendix A) is equal to

$$\frac{J_0(T/2)\lambda(T/2)}{J(T/2)}, \tag{2.110}$$

where $J_0(\tau)$ and $J(\tau)$, which are called Jacobi fields, satisfy the following equations

$$\left[-\frac{d^2}{d\tau^2}+\omega^2\right]J_0(\tau) = 0 \tag{2.111}$$

and

$$\left[-\frac{d^2}{d\tau^2} + V''(\bar{x})\right] J(\tau) = 0 \tag{2.112}$$

under the initial conditions

$$J_0(-T/2) = J(-T/2) = 0 \tag{2.113}$$

and

$$J_0'(-T/2) = J'(-T/2) = 1. \tag{2.114}$$

From Equation (2.111) we immediately find at $\tau \to T/2 \to \infty$

$$J_0(\tau \to T/2) \to \frac{1}{2\omega} \exp[\omega T]. \tag{2.115}$$

As for Equation (2.112), the time derivative of the instanton solution $\eta(\tau) \equiv d\bar{x}(\tau)/d\tau$ is the solution of this equation. If the instanton trajectory is chosen as $\bar{x}(\tau) = -\bar{x}(-\tau)$, then $\eta(\tau)$ is symmetric and asymptotically given as

$$\eta(\tau) \overset{\tau \to \pm\infty}{\longrightarrow} P \exp[-\omega|\tau|], \tag{2.116}$$

where the constant P will be found later. Denoting the second independent solution that satisfies the Wronskian $W(\xi, \eta) = \eta\xi' - \eta'\xi = 1$ as $\xi(\tau)$, we have

$$J(\tau) = -\xi(-T/2)\eta(\tau) + \xi(\tau)\eta(-T/2) \tag{2.117}$$

and

$$\xi(\tau) \overset{\tau \to \pm\infty}{\longrightarrow} \frac{1}{2\omega P} \exp[\omega|\tau|]. \tag{2.118}$$

Thus we find at $\tau = T/2$

$$J(T/2) = \frac{1}{\omega} + O\left(\exp\left[-\frac{\omega}{2}T\right]\right) \tag{2.119}$$

and

$$\text{Equation (2.110)} = \frac{\exp[\omega T]}{2} \lambda(T/2). \tag{2.120}$$

The lowest eigenvalue $\lambda(T/2)$ is obtained as follows. We consider the eigenvalue problem

$$-\frac{d^2\Psi}{d\tau^2} + V''(\tau)\Psi = \lambda\Psi \tag{2.121}$$

with the boundary conditions

$$\Psi\left(\pm\frac{T}{2}\right) = 0. \tag{2.122}$$

The solution can be formally written as

$$\Psi = \Psi_0 + \lambda \int_{-T/2}^{\tau} d\tau' [\eta(\tau)\xi(\tau') - \eta(\tau')\xi(\tau)]\Psi(\tau'), \tag{2.123}$$

where Ψ_0 is the solution for $\lambda = 0$. Since $\Psi_0(\tau)$ satisfies the zero boundary condition $\Psi_0(-T/2) = 0$ and the normalization is not necessary, we can put

$$\Psi_0(\tau) = \eta(\tau) + C\xi(\tau). \tag{2.124}$$

From $\Psi_0(-T/2) = 0$ and the asymptotic behavior of ξ and η, C can be obtained as

$$C = 2\omega P^2 \exp[-\omega T]. \tag{2.125}$$

The second boundary condition $\Psi(T/2) = 0$ gives

$$\lambda = -\frac{\Psi_0(T/2)}{\int_{-T/2}^{T/2} d\tau'[\eta(T/2)\xi(\tau') - \eta(\tau')\xi(T/2)]\Psi(\tau')}. \tag{2.126}$$

Replacing $\Psi(\tau')$ in the denominator by $\Psi_0(\tau')$, we obtain the desired λ as

$$\lambda(T/2) = \frac{4\omega P^2}{S_0} \exp[-\omega T]. \tag{2.127}$$

Here we have used Equations (2.116), (2.118), (2.125), and

$$\int_{-T/2}^{T/2} \eta^2(\tau) d\tau = \int_{-T/2}^{T/2} \left(\frac{d\bar{x}}{d\tau}\right)^2 d\tau = S_0 \text{ [see Equation (2.94)]}. \tag{2.128}$$

Using Equations (2.120) and (2.127), we have the following expression for the pre-exponential factor B [see Equation (2.108)]:

$$B = 2P\sqrt{\frac{\omega}{\pi\hbar}}. \tag{2.129}$$

Finally, the constant P is obtained from the classical equation of motion

$$\tau = \int^x \frac{dz}{p(z)} = \int_{-x_0}^x dz \left(\frac{1}{p} - \frac{1}{\omega(z + x_0)}\right) + \frac{1}{\omega} \ln(x + x_0) + C_0, \tag{2.130}$$

where $p(z) = \sqrt{2V(z)}$. The constant of integration C_0 is determined from the condition $x(0) = 0$ as

$$\exp[-\omega C_0] = x_0 \exp\left[\int_{-x_0}^0 dz \left(\frac{\omega}{p} - \frac{1}{z + x_0}\right)\right]. \tag{2.131}$$

Equation (2.130) gives the instanton trajectory $\bar{x}(\tau)$ and $\eta(\tau)$ is obtained as its derivative ($= d\bar{x}/d\tau$). The asymptotic expression at $\tau \to -T/2$ of thus obtained $\eta(\tau)$ can be compared with Equation (2.116) and we obtain the following expression of P:

$$P = \omega \exp[-\omega C_0] = \omega x_0 \exp\left[\int_{-x_0}^0 dz \left[\frac{\omega}{p} - \frac{1}{z + x_0}\right]\right]. \tag{2.132}$$

It is more convenient to rewrite the above integral in terms of τ. Using the relation

$$\int_{-x_0}^{x} dz \left(\frac{\omega}{p(z)} - \frac{1}{z + x_0} \right) = \ln \left(\frac{p(x)}{\omega(x + x_0)} \right) + \int_{-\infty}^{\tau} d\tau \left(\omega - \frac{dp}{dz} \right), \tag{2.133}$$

we can have

$$P = p(0) \exp \left[\int_{-\infty}^{0} d\tau \left(\omega - \frac{dp}{dz} \right) \right]. \tag{2.134}$$

Then the pre-exponential factor B can be finally obtained as

$$B = \sqrt{\frac{4\omega}{\pi\hbar}} p(0) \exp \left[\int_{-\infty}^{0} d\tau \left(\omega - \frac{dp}{dz} \right) \right]. \tag{2.135}$$

So far we have been using the unit mass, but now we can recover the mass in this final expression. From the dimension analysis we obtain finally the ground-state tunneling splitting [see Equation (2.102)] as

$$\Delta_0 = \hbar\omega \sqrt{\frac{4p^2(0)}{m\pi\omega\hbar}} \exp \left[\int_{-\infty}^{0} d\tau \left(\omega - \frac{1}{m} \frac{dp}{dz} \right) \right] \exp[-S_0/\hbar]. \tag{2.136}$$

Now we apply the above expressions to the quartic potential problem,

$$V(x) = \lambda(x^2 - x_0^2)^2. \tag{2.137}$$

The instanton trajectory, the corresponding momentum, and the frequency ω can be obtained as

$$x(\tau) = x_0 \tanh(\omega\tau/2), \tag{2.138}$$

$$p(x) = \sqrt{2\lambda}(x_0^2 - x^2) = \sqrt{2\lambda} x_0^2 \operatorname{sech}^2(\omega\tau/2), \tag{2.139}$$

and

$$\omega = \sqrt{8x_0^2\lambda}. \tag{2.140}$$

Then we finally obtain the well-known expression of the splitting as

$$\Delta_0 = \hbar\omega \sqrt{\frac{2m\omega^3}{\pi\lambda\hbar}} \exp \left[-\frac{m\omega^3}{12\lambda\hbar} \right]. \tag{2.141}$$

Next, let us consider the decay of a metastable state through tunneling [see Figure 1.1(b) and Figure 2.10]. The formulation can be done in the similar way as in the case of tunneling splitting [44]. The decay rate is given by

$$k = B \exp[-S_0/\hbar], \tag{2.142}$$

where the pre-exponential factor B can be obtained as

$$B = \sqrt{\frac{S_0}{2\pi\hbar}} \left[\frac{\det'\left(-\frac{d^2}{d\tau^2} + V''(\tau)\right)}{\det\left(-\frac{d^2}{d\tau^2} + \omega^2\right)} \right]^{-1/2}. \tag{2.143}$$

It should be noted that the pre-exponential factor is one-half of the tunneling splitting case [see Equation (2.108)] and the action S_0 is the action along the instanton from the potential minimum at $\tau = -\infty$ back to the potential minimum at $\tau = \infty$ and thus is two times of the action from the potential minimum to the turning point (= exit to the free space). By using the same procedure as that in the tunneling splitting, the decay rate k is finally obtained as

$$k = \omega\sqrt{\frac{m\omega x_0^2}{\pi\hbar}} \exp\left[\int_0^{x_0} dx\left(\frac{m\omega}{p} - \frac{1}{x}\right)\right] \exp[-S_0/\hbar], \tag{2.144}$$

where $x = 0(\tau = -\infty)$ and $x = x_0(\tau = 0)$ correspond to the potential minimum and the turning point (exit), respectively.

In the case of cubic potential

$$V(x) = \frac{1}{2}m\omega^2 x^2\left(1 - \frac{x}{x_0}\right), \tag{2.145}$$

we obtain

$$S_0 = \frac{8}{15}m\omega x_0^2 \tag{2.146}$$

and

$$k = 4\omega\left(\frac{m\omega x_0^2}{\pi\hbar}\right)^{1/2} \exp[-S_0/\hbar]. \tag{2.147}$$

2.5.2 Modified WKB Method

In this subsection, we show that the same results as those obtained by the instanton theory in the previous subsection can be derived by the WKB theory with a slight modification [43,46]. We consider the tunneling splitting in a symmetric double well potential and as usual we use the asymptotic WKB wave function localized in one of the wells—say, left-side well:

$$\Psi = \exp\left[-\frac{W_0}{\hbar} - W_1\right]. \tag{2.148}$$

The key observation here is that the energy E_0 of the ground state is given by

$$E_0 = \hbar\frac{\omega}{2} + O(\hbar^2), \tag{2.149}$$

and we treat this energy as the first-order term with respect to $\hbar$ [53]. Then W_0 and W_1 satisfy the following forms of Hamilton-Jacobi equation and transport equation (or continuity equation),

$$\frac{1}{2}\left(\frac{dW_0}{dx}\right)^2 - V(x) = 0 \tag{2.150}$$

and

$$\frac{dW_1}{dx}\frac{dW_0}{dx} - \frac{1}{2}\frac{d^2W_0}{dx^2} + \frac{\omega}{2} = 0. \tag{2.151}$$

These can be solved easily as

$$W_0(x) = \int_{-x_0}^{x} p(z)dz \tag{2.152}$$

and

$$W_1(x) = \frac{1}{2}\ln[p(x)] - \frac{1}{2}\int^{x} \frac{\omega}{p(z)}dz. \tag{2.153}$$

It should be noted that the energy is missing in Equation (2.150) but appears in Equation (2.151). The integral $\int dz/p(z)$ is regularized, or the divergence is removed, in the same way as before and we come to the solution

$$\Psi(x) = \sqrt{\frac{x+x_0}{p(x)}}\exp\left(\int_{-x_0}^{x}\left[\frac{\omega}{p(z)} - \frac{1}{z+x_0}\right]dz\right)\exp\left[-\int_{-x_0}^{x} p(z)dz\right]. \tag{2.154}$$

$p(x)$ can be thought of as the classical momentum, and the corresponding (instanton) trajectory $x(\tau)$ runs on the inverted potential $-V(x)$, starting from $-x_0$ with zero velocity. This trajectory is assumed to be symmetric in time and passes the barrier top at $\tau = 0$. At x close to $-x_0$, we have $V(x) \simeq \omega^2(x+x_0)^2/2$, $p(x) \simeq \omega(x+x_0)$, $W_0(x) = \omega(x+x_0)^2/2$ and $W_1 = 0$. Thus the wave function Equation (2.148) within the accuracy of order $O(\hbar^2)$ becomes at $x \simeq -x_0$

$$\Psi(x) = \exp\left[-\frac{\omega(x+x_0)^2}{2\hbar}\right]. \tag{2.155}$$

This semiclassical wave function does not require any special matching condition at the boundary between classically allowed and classically forbidden regions, which represents the main obstacle in the ordinary WKB theory. The tunneling splitting Δ_0 can be calculated from the Herring formula [54],

$$\Delta_0 = -\hbar^2\frac{2\Psi(x)\frac{d\Psi(x)}{dx}|_{x=0}}{\int \Psi^2(x)dx}. \tag{2.156}$$

The denominator can be easily calculated using Equation (2.155). In the numerator, within the accuracy of semiclassical approximation, the differentiation of $\Psi(x)$ at $x = 0$ gives

$$\frac{\Psi(x)}{dx}|_{x=0} = -\frac{1}{\hbar}\frac{dW_0}{dx}|_{x=0} = -\frac{p(x=0)}{\hbar}. \tag{2.157}$$

Transforming the integral of $\omega/p - 1/(z+x_0)$ with respect to z in Equation (2.154) to the integral with respect to τ as before using Equation (2.133) and introducing the mass m, we obtain the same formula as Equation (2.136).

The formulation here indicates that the WKB theory works well and gives the same results as the instanton theory, if we observe that the energy is the quantity proportional to $\hbar$. This modified WKB method is actually simpler than the instanton treatment and can be directly extended to multidimensional systems. This is discussed in Chapters 6 and 7.

2.6 ENERGY LEVELS IN A DOUBLE WELL POTENTIAL

2.6.1 Asymmetric Double Well Potential

Here we consider the energy levels or the energy shifts in an asymmetric double well potential such as that shown in Figure 2.6. The diagrammatic technique mentioned in Section 2.4 can be applied. The corresponding diagram is shown in Figure 2.7.

As was explained in Section 2.4, M and $x_\alpha(\alpha = a-d)$ represent the 2×2 tunneling transfer matrix and turning points. The matrix M is given by Equations (2.24), (2.25), and (2.69). The secular equation is given by

$$\sum_j [Q]_{1j} = \sum_j [Q]_{2j} \tag{2.158}$$

with

$$Q = L(x_c, x_d)ML(x_a, x_b), \tag{2.159}$$

where

$$[L(a, b)]_{ij} = \delta_{ij} \exp[i\sigma(a, b) + i\pi/4] \tag{2.160}$$

and

$$\sigma(a, b) = \int_a^b K(x)dx = \sqrt{\frac{2m}{\hbar^2}} \int_a^b \sqrt{E - V(x)}dx. \tag{2.161}$$

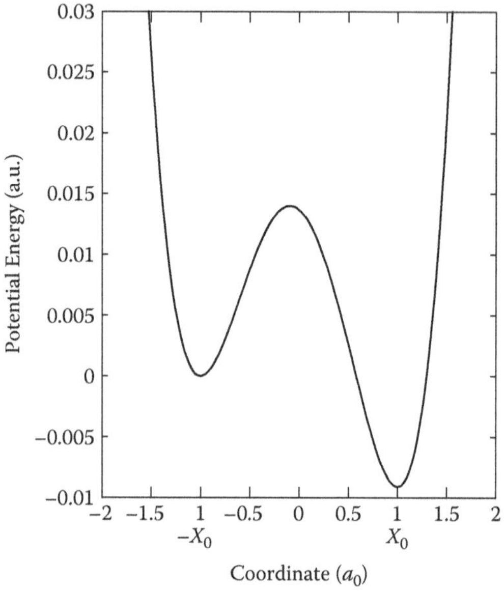

FIGURE 2.6 Asymmetric double well potential.

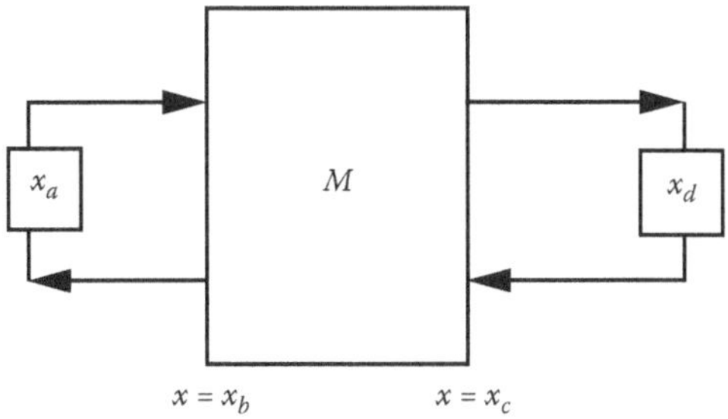

FIGURE 2.7 Diagram corresponding to Figure 2.6.

Equation (2.158) gives

$$F \equiv |M_{11}| \sin[\sigma(x_a, x_b) + \sigma(x_c, x_d) + \pi/2 - \phi(\epsilon)]$$
$$= -|M_{12}| \sin[\sigma(x_c, x_d) - \sigma(x_a, x_b) + \pi/2] \equiv G. \quad (2.162)$$

Here $\phi(\epsilon)$ is given by Equation (2.25) and

$$|M_{11}| = \sqrt{1+\kappa^2}, \quad |M_{12}| = \kappa, \quad (2.163)$$

$$\kappa = \exp[-\pi\epsilon], \quad (2.164)$$

$$\epsilon = -\frac{1}{\pi}\int_{x_b}^{x_c} |K(x)|dx \text{ [see Equation (2.69)]}. \quad (2.165)$$

The eigenvalues can be obtained from the crossing points of the two curves, $F(E)$ and $G(E)$. As an example the following potential is considered:

$$V(x) = V_0(x^2 - x_0^2)^2 + AV_0(x^2 - x_0^2)x - 2AV_0(x + x_0)x_0^2, \quad (2.166)$$

where $V(-x_0) = 0$. The parameters used are $x_0 = 1.0a_0$, $m = 1000$ a.u., $V_0 = 4000\,\mathrm{cm}^{-1}/a_0^4 = 1.823 \times 10^{-2}$ a.u., and $A = 0.125$ a.u. The potential function $V(x)$ is shown in Figure 2.6.

The functions $F(E)$ and $G(E)$ can be easily calculated and plotted as a function of E to find crossing points. An example is shown in Figure 2.8. The accurate numerical solutions give the lowest three eigenvalues as $E_{0-2} = -3.04 \times 10^{-3}$, 5.41×10^{-3}, and 7.90×10^{-3} a.u. The lowest one is the level in the right-side single well. The two crossing points in Figure 2.8 correspond to the second- and third-lowest eigenvalues.

In the case that the unperturbed eigenvalue, say E_l, of the left-side well is close to the one (E_r) in the right-side well, Miller [17,55] obtained the formula for the level shift of E_l by using the spectral function method

$$\Delta E = \frac{\exp[-2\pi|\epsilon|]}{4\pi^2 \frac{dn_1(E_r)}{dE}\frac{dn_2(E_l)}{dE}(E_r - E_l)}, \quad (2.167)$$

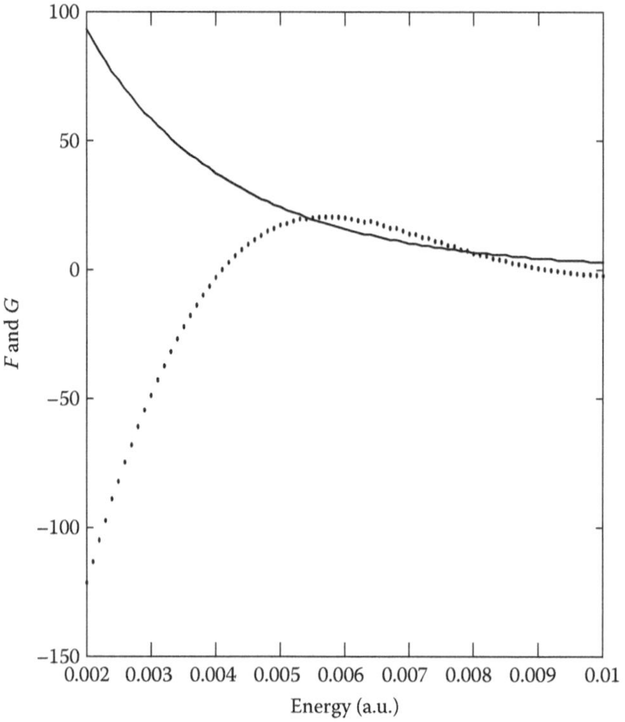

FIGURE 2.8 The functions $F(E)$ (dotted line) and $G(E)$ (solid line) [see Equation (2.162)].

where

$$n_1(E) = \frac{1}{\pi}\int_{x_a}^{x_b} K(x)dx - 1/2. \tag{2.168}$$

2.6.2 Symmetric Double Well Potential

This case was already treated by the instanton theory and the modified WKB theory in Section 2.5. Here we apply these methods to a model potential shown in Figure 2.9. This is the one given by Equation (2.166) with the parameter $A = 0$. The accurate numerical solution gives the lowest two eigenvalues as $E_0 = 5.7444 \times 10^{-3}$ a.u. and $E_1 = 5.7706 \times 10^{-3}$ a.u. with the splitting $\Delta_0 = 2.62 \times 10^{-5}$ a.u. $= 5.75$ cm^{-1}. The instanton theory and the modified WKB theory explained in the previous section [see Equation (2.141)] give a good estimate as $\Delta_0 = 6.63$ cm^{-1}. The diagrammatic technique can in principle be applied to this case to obtain eigenvalues but is actually not efficient. As can be easily guessed, it is better to directly evaluate the splitting.

Using the spectral function method, Miller [17,55] obtained the following expression for the tunneling splitting, which is applicable to the excited state E_n:

$$\Delta_n = \frac{\exp[-\pi|\epsilon(E_n)|]}{\pi dn(E_n)/dE}. \tag{2.169}$$

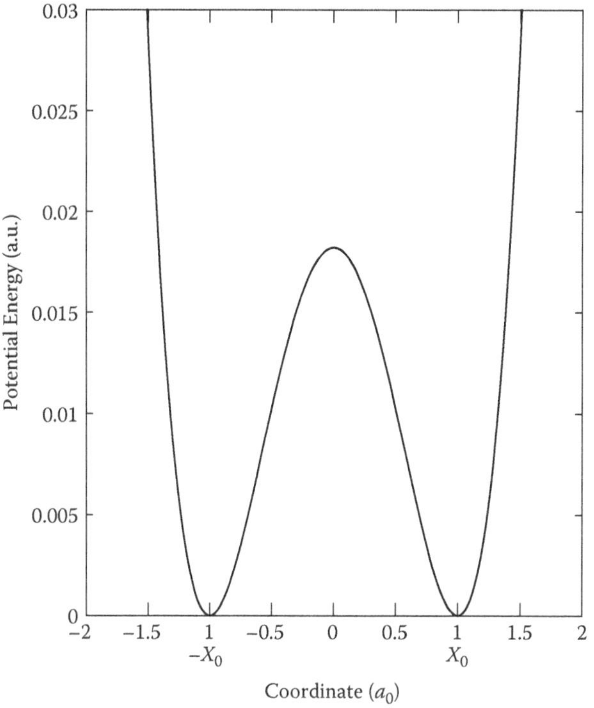

FIGURE 2.9 Symmetric double well potential.

The modified WKB method can also be applied to low excited states. This is discussed for the case of a multidimensional problem in Chapter 6.

2.7 DECAY OF METASTABLE STATE

A potential corresponding to this problem is Figure 1.1(b). Metastable states supported by the potential well decay through tunneling. The corresponding diagram is shown in Figure 2.10.

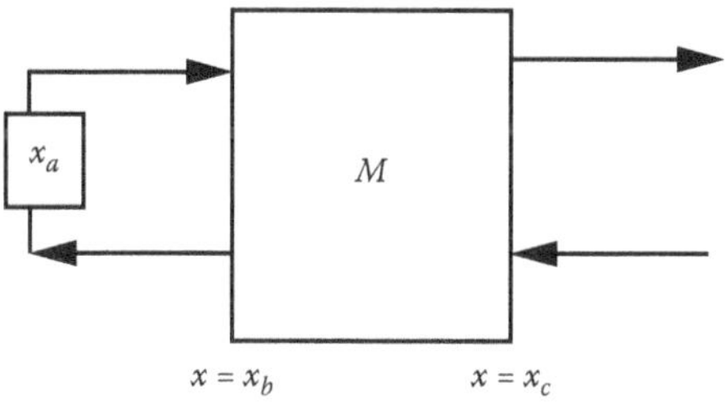

FIGURE 2.10 Diagram corresponding to Figure 1.1(b).

This problem can be treated as a resonant scattering, and the scattering matrix S or the elastic scattering phase shift η are given by

$$S \equiv \exp[2i\eta] = \frac{C_1}{C_2} \tag{2.170}$$

with

$$\begin{pmatrix} C_1 \\ C_2 \end{pmatrix} = \begin{pmatrix} e^{i\eta_0} & 0 \\ 0 & e^{-i\eta_0} \end{pmatrix} ML(x_a, x_b) \begin{pmatrix} 1 \\ 1 \end{pmatrix}, \tag{2.171}$$

where η_0 is the elastic scattering phase shift by the potential at $x \geq x_c$ and given by

$$\eta_0 = \int_{x_c}^{\infty} (K(x) - K(\infty))\, dx - K(\infty)x_c. \tag{2.172}$$

Then we have

$$\eta = \eta_0 + \frac{\pi}{4} + \frac{\phi}{2} + \gamma, \tag{2.173}$$

where

$$\tan\gamma = \frac{\sqrt{1+\kappa^2} - \kappa}{\sqrt{1+\kappa^2} + \kappa} \tan\alpha, \tag{2.174}$$

with

$$\alpha = \sigma(x_a, x_b) + \phi. \tag{2.175}$$

If the complex resonance energy is denoted as $E_r - i\Gamma/2$, then η near the resonance position E_r is given by

$$\eta \simeq \eta_{bg} + \tan^{-1}\left[\frac{\Gamma/2}{E - E_r}\right], \tag{2.176}$$

where η_{bg} represents the background scattering phase shift. If we expand α around the resonance energy E_r as

$$\alpha(E) \simeq \alpha(E_r) + \left[\frac{d\alpha}{dE}\right]_{E_r} (E - E_r) + \cdots, \tag{2.177}$$

then the resonance position E_r and its width Γ can be obtained as

$$\alpha(E_r) = \int_{x_a}^{x_b} K(x)dx + \frac{\phi}{2} = \left(n + \frac{1}{2}\right)\pi \quad (n = 0, 1, 2, \ldots) \tag{2.178}$$

$$\Gamma = 2\frac{\sqrt{1+\kappa^2} - \kappa}{\sqrt{1+\kappa^2} + \kappa}\left[\frac{d\alpha}{dE}\right]^{-1/2} |_{E_r} = 4\frac{\sqrt{1+\kappa_r^2} - \kappa_r}{\sqrt{1+\kappa_r^2} + \kappa_r} \Big/ \int_{x_a}^{x_b} \frac{dx}{K(x)}, \tag{2.179}$$

where $\kappa_r \equiv \kappa(E_r)$. The decay rate k of the metastable (resonance) state is equal to $\Gamma/\hbar$. The scattering probability is proportional to $\sin^2\eta$, which is given by

$$\sin^2\eta(E) = \sin^2\eta_0(E) + \frac{(\Gamma/2)^2\cos(2\eta_0) + (\Gamma/2)(E_r - E)\sin(2\eta_0)}{(E - E_r)^2 + (\Gamma/2)^2}. \tag{2.180}$$

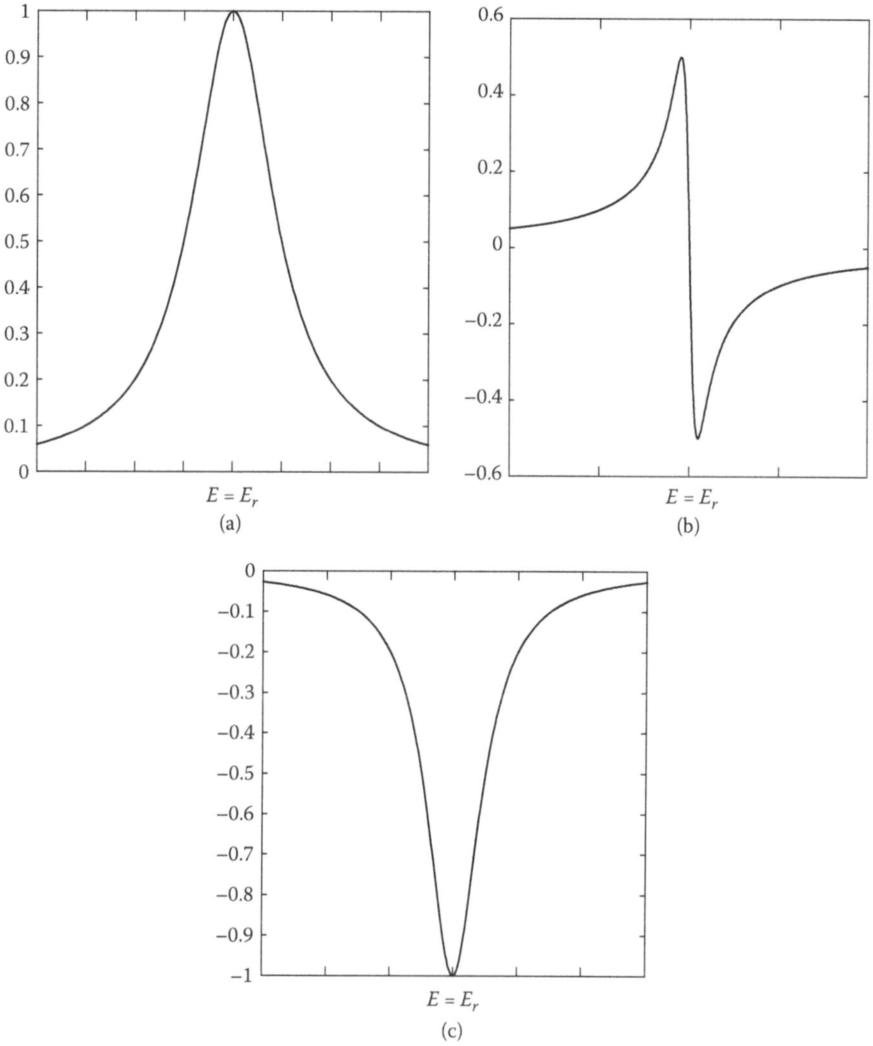

FIGURE 2.11 Shape of resonance peak, $\sin^2 \eta$, against energy E. (a) $\eta_0 = 0$, (b) $\eta_0 = \pi/4$, (c) $\eta_0 = \pi/2$.

Depending on the value of η_0, the curve of $\sin^2 \eta(E)$ vs. E changes from Lorentzian peak to dip. The cases of $\eta_0 = 0, \pi/4$, and $\pi/2$ are schematically shown in Figures 2.11(a)–(c), respectively.

The WKB treatment and comparison with the instanton theory made by many authors are discussed in the book by Benderskii et al. [17] (see also [52]). We don't go into the details here.

3 Two-Dimensional Theory

3.1 WKB THEORY

It is rather obvious that one-dimensional theory is not good enough to comprehend quantum mechanical tunneling phenomena. For instance, the tunneling splitting estimated using the one-dimensional potential along the intrinsic reaction coordinate gives a few orders of magnitude smaller value [56]. There is a more crucial example that demonstrates the conceptual deficiency of one-dimensional treatment and shows interesting features peculiar to multidimensional tunneling phenomena. This is explained in Chapter 4. In this sense, multidimensional treatment is definitely needed. Here we describe the two-dimensional WKB theory developed by Takada and Nakamura [30].

There are two fundamental problems to be solved in developing any multidimensional theory: (1) connection of wave functions between the classically allowed and forbidden regions, and (2) wave propagation in the classically forbidden region. The first question is not usually paid proper attention, simply because it is not an easy subject. However, the second problem strongly depends on the character of the wave function in the classically allowed region, namely, on problem (1). So, we start from problem (1), confining ourselves to the case where classical motion is not chaotic and to the problem of tunneling splitting in a symmetric double well potential. Hereafter the particle mass is assumed to be unity.

The first task is to construct semiclassical wave function in the classically allowed region. As is well known, the KAM torus exists according to the Kolmogorov-Arnold-Moser (KAM) [57,58] and this integral system can be quantized by the Einstein-Brillouin-Keller (EBK) quantization rule [58] as

$$I_j \equiv \oint_{C_j} \mathbf{p} \cdot d\mathbf{x} = 2\pi \hbar (n_j + \beta_j/4), \quad j = 1, 2, \tag{3.1}$$

where $\mathbf{p}$ is momentum, n_j is a quantum number corresponding to the action, C_j is an irreducible circle on the KAM torus, and β_j is the Maslov index. In the case of vibrational motion, β_j is equal to 2 and it is 0 in the case of rotational motion. The eigenvalue $E_\mathbf{n}$ of a quantized state is equal to the Hamiltonian H expressed in terms of the actions I_j's as

$$E_\mathbf{n} \equiv E_{n_1,n_2} = H(I_1, I_2). \tag{3.2}$$

The semiclassical wave function for the EBK quantized state has been formulated by Malsov and co-workers [59,60]. They gave a prescription to avoid the divergence of the primitive semiclassical wave function at caustics. This is briefly explained here for later convenience by taking an example of two-dimensional nonlinear coupled oscillators [61]. Figure 3.1 illustrates a configuration space in which the projection of the KAM torus becomes a distorted rectangular. In the inner region (I) (see Figure 3.1)

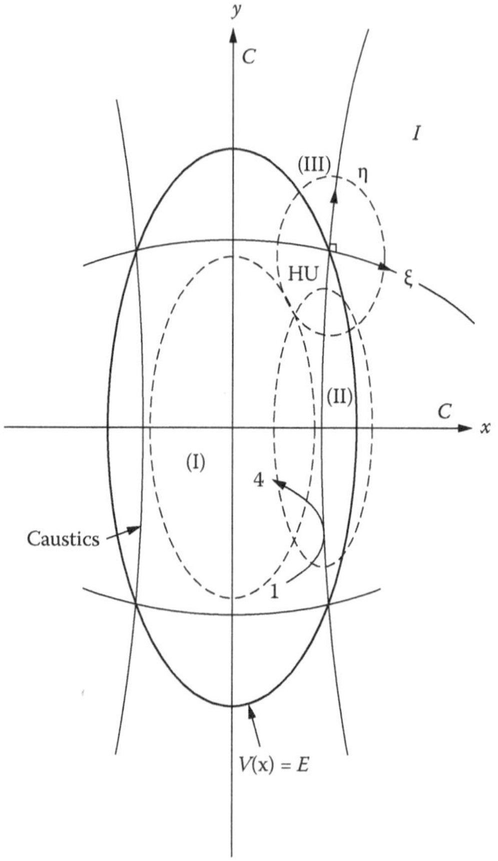

FIGURE 3.1 Coordinate space (x, y) near one of the potential wells. Distorted rectangle represents the KAM torus projected on this space. (I), (II), and (III) specify the regions to be treated separately in Maslov's theory. Tunneling region is divided into two kinds: C and I. A typical trajectory that reflects back at the right caustics is depicted with the branch numbers 1 and 4. (Taken from Reference [30] with permission.)

the primitive semiclassical wave function is nonsingular and given by

$$\Psi_{\mathbf{n}} = \sum_{\nu=1}^{4} (\rho_\nu)^{1/2} \exp\left(\frac{i}{\hbar} W_\nu\right), \tag{3.3}$$

where $W_\nu(x, y, \mathbf{I})$ and $\rho_\nu(x, y, \mathbf{I})$ are the solutions of the Hamilton-Jacobi equation

$$\frac{1}{2}(\nabla W_\nu)^2 + V(\mathbf{x}) = E_{\mathbf{n}} \tag{3.4}$$

and the transport (or continuity) equation

$$\nabla \cdot (\rho_\nu \nabla W_\nu) = 0. \tag{3.5}$$

It should be noted that the energy $E_\mathbf{n}$ is retained in Equation (3.4) as usual in the ordinary WKB theory. Although the action W and, accordingly, the density ρ are single-valued functions on the torus in phase space, they become multivalued after being projected onto configuration space. Subscript $\nu(= 1, 2, 3, 4)$ classifies trajectories according to the sign of the momenta $p_x = \partial W/\partial x$ and $p_y = \partial W/\partial y$. The branches $\nu = 1, 2, 3$, and 4 correspond to $(p_x, p_y) = (+, +), (+, -), (-, -)$, and $(-, +)$, respectively. The Hamilton-Jacobi equation can be solved by the method of characteristics [62]. The solution W_ν can be solved as

$$W_\nu(\mathbf{x}) = \int^{\mathbf{x}} \mathbf{p}_\nu(\mathbf{x}) \cdot d\mathbf{x} \tag{3.6}$$

by using the classical trajectories on the quantized torus. In region (II) of Figure 3.1, which contains the caustics in the classically accessible region, the wave function Equation (3.3) cannot be utilized because of its divergence at the caustics. If we use the representation in the mixed space (p_x, y), this divergence can be avoided. The action and density are given by the solutions of the Hamilton-Jacobi and the continuity equations in this mixed space. The corresponding wave function $\Psi_\mathbf{n}(x, y)$ in the coordinate space can be obtained by Fourier transformation. In this way the wave functions in all the regions (I)–(III) containing the caustics (see Figure 3.1) are obtained. These are called Lagrange manifold [59,60]. Finally, these wave functions are matched in each overlapping region to form a global uniform semiclassical wave function. This matching can be simply summarized as the simple statement that the wave function Equation (3.3) loses its phase by $\pi\hbar/2$ each time when the trajectory touches the caustics. For instance, in the case of the trajectory shown in Figure 3.1 we have $W_4 = W_1 - \pi\hbar/2$. We can also find simple relations, $\mathbf{p}_1 = \nabla W_1 = \nabla W_4 = -\nabla W_2 = -\nabla W_3$ at the right-side caustics. Thus we finally have Table 3.1.

Let us move into the tunneling region near the hyperbolic umbilic (HU) point (see Figure 3.1). Near the hyperbolic umbilic point the wave function is given by the Airy function as

$$\Psi_\mathbf{n}(\mathbf{x}) \propto \mathrm{Ai}(\alpha\xi)\mathrm{Ai}(\beta\eta), \tag{3.7}$$

where $\mathrm{Ai}(x)$ is the Airy function, (ξ, η) is the local coordinate system (see Figure 3.1), and the parameter $\alpha\,(\beta)$ represents the potential gradient in the $\xi\,(\eta)$ direction

TABLE 3.1
Interrelations among W_ν's on the Caustics

Caustics	Relations
Right	$W_1 = -W_2 + \pi\hbar/2 = -W_3 = W_4 + \pi\hbar/2$
Left	$W_1 = -W_2 - \pi\hbar/2 = -W_3 = W_4 - \pi\hbar/2$
Top	$W_1 = W_2 + \pi\hbar/2 = -W_3 = -W_4 + \pi\hbar/2$
Bottom	$W_1 = W_2 - \pi\hbar/2 = -W_3 = -W_4 - \pi\hbar/2$

(respectively). With use of the asymptotic expansions of the Airy function

$$\mathrm{Ai}(x) \to \begin{cases} \frac{1}{2}\pi^{-1/2}x^{-1/4}\exp\left[-\frac{2}{3}x^{3/2}\right], & \text{for } x \gg 1, \\ \pi^{-1/2}|x|^{-1/4}\sin\left(\frac{2}{3}|x|^{3/2}+\pi/4\right), & \text{for } x \ll -1, \end{cases} \tag{3.8}$$

the asymptotic behavior of the wave function Equation (3.7) in the classically forbidden region of $\xi > 0$ and $\eta < 0$ can be obtained, to exponential accuracy, as

$$\Psi_{\mathbf{n}} \propto \exp\left[-\frac{2}{3}(\alpha\xi)^{3/2}\right]\exp\left[i\frac{2}{3}|\beta\eta|^{3/2}+i\pi/4\right]+\text{c.c.} \tag{3.9}$$

In comparison with Equation (3.3) the action W becomes complex,

$$W_{\pm} = \pm W_R(\eta) + iW_I(\xi), \tag{3.10}$$

where the subscripts $\pm$ specify the branch mentioned above. On the other hand, in the first quadrant $\xi, \eta > 0$ the wave function Equation (3.7) becomes

$$\Psi_{\mathbf{n}} \propto \exp\left[-\frac{2}{3}(\alpha\xi)^{3/2}-\frac{2}{3}(\beta\eta)^{3/2}\right], \tag{3.11}$$

which indicates that the action is pure imaginary,

$$W(\mathbf{x}) = iW_I(\xi,\eta). \tag{3.12}$$

Thus there are three distinct regions around the hyperbolic umbilic (HU) point: The classically allowed R-region, the classically forbidden C-region, and the classically forbidden I-region in which the corresponding action is real, complex, and pure imaginary, respectively [29]. The four branches in R-region connect to two branches in C-region and the latter connects to one branch of I-region. The wave function in the C-region can have nodal lines. The boundaries between the C- and I-regions form caustics in the classically forbidden region. Wave propagation in the C-region is given by the following Hamilton-Jacobi equation by inserting Equation (3.10) into Equation (3.4),

$$(\nabla W_R)^2 - (\nabla W_I)^2 = 2[E_{\mathbf{n}} - V(x)] \tag{3.13}$$

and

$$\nabla W_R \cdot \nabla W_I = 0. \tag{3.14}$$

The conventional classical trajectory technique cannot be used to solve these equations. Instead, the Huygens-type wave propagation should be employed [7]. Its essential idea is to propagate the W_I equisurface step by step. We don't go into details here [30]. On the other hand, wave propagation in the I-region can be made by the well-known equation

$$\frac{1}{2}(\nabla W_I)^2 - V(x) = -E_{\mathbf{n}}. \tag{3.15}$$

This equation can be solved by using classical trajectory on the inverted potential.

The final important task is to connect the wave functions between different regions. This can be carried out by assuming local separability—namely, the semiclassical connection formulas in one dimension given in Section 2.2. As an example, we consider here the connection from R-region to C-region at the right-side caustics and then from this C-region to the upper I-region (see Figure 3.1). According to Table 3.1, the wave function near the right caustics in the R-region is given by

$$\Psi_{\mathbf{n}}(\xi,\eta) = \frac{4c}{\sqrt{p_\xi p_\eta}} \cos\left(\frac{W_1(0,\eta)}{\hbar} - \frac{\pi}{4}\right) \cos\left(\frac{1}{\hbar}\int_\xi^0 |p_\xi| d\xi - \frac{\pi}{4}\right). \tag{3.16}$$

Applying the one-dimensional connection formula in the ξ direction, we obtain the wave function in the C-region as

$$\Psi_{\mathbf{n}} = \frac{2c}{\sqrt{p_R p_I}} \cos(W_R(\eta)/\hbar)\exp[-W_I(\xi)/\hbar] \tag{3.17}$$

with $p_R = p_\eta$ and $p_I = \partial W_I/\partial\xi$, where

$$W_R(\eta) = W_1(0,\eta) - \pi\hbar/4 \tag{3.18}$$

and

$$W_I(\xi) = \int_0^\xi d\xi' \sqrt{2[V(\xi',\eta) - V(0,\eta)]}. \tag{3.19}$$

The connection from the C-region to the I-region can be carried out in the same way. The wave function in the I-region near the caustics is given by

$$\Psi_{\mathbf{n}}(\xi,\eta) = \frac{c'}{\sqrt{p_\xi p_\eta}} \exp\left(-\frac{1}{\hbar}[W_{I_\eta}(\eta) + W_{I_\xi}(\xi)]\right), \tag{3.20}$$

where

$$W_{I_\xi}(\xi) = W_I(\xi), \tag{3.21}$$

$$W_{I_\eta}(\eta) = \int_0^\eta d\eta' \sqrt{2[V(\xi,\eta') - V(\xi,0)]}, \tag{3.22}$$

$$p_\eta = \frac{\partial W_{I_\eta}}{\partial\eta} \tag{3.23}$$

and $p_\xi = p_I$. Equation (3.23) determines the initial condition of classical trajectory in the I-region. Since $p_\eta = 0$ at the boundary, the trajectory leaves tangentially to the caustics. In these equations ξ is defined as the arc length along the caustics measured from the hyperbolic umbilic (HU) point and η is a perpendicular deviation from the caustics.

Now the physical picture of multidimensional tunneling obtained from the above analysis is explained by taking a symmetric mode coupling model potential [30]. The Hamiltonian is given by

$$H = -\frac{g^2}{2}\left(\frac{\partial^2}{\partial x^2} + \frac{\partial^2}{\partial y^2}\right) + \frac{1}{8}(x+1)^2(x-1)^2 + \frac{\omega_y^2}{2}\left(y + \frac{\gamma}{\omega_y^2}(x^2-1)\right)^2, \tag{3.24}$$

where all quantities are scaled to be dimensionless. The parameters g, ω_y, and γ play the role of the Planck constant, relative frequency in the y direction, and coupling strength, respectively. This model potential is widely used to represent proton tunneling [4,35,56]. For instance, in the case of malonaldehyde the coordinate x represents the motion of the hydrogen atom transferring from one oxygen atom to the other oxygen atom, while y roughly represents the scissors-like motion of the O–C–C–C–O frame. Figure 3.2 shows typical tunneling pictures.

The most typical one is Figure 3.2(a), which is the case of modest values of coupling strength γ and frequency ω_y, representing the tunneling mainly through the I-region. Starting from the R-region, tunneling proceeds in the C_1-region for a while and the trajectory taking off from this region at a certain position $\mathbf{x}_0$ mainly contributes to the tunneling splitting. Based on this picture, we can formulate an analytical expression of tunneling splitting. Basic assumptions are as follows. (1) The potential is assumed to be separable in between the hyperbolic umbilic point and the taking-off point $\mathbf{x}_0$. Thus the caustics is extended into the classically forbidden region as a straight line along the η axis. (2) The potential is assumed to be linear near the hyperbolic umbilic point and $\mathbf{x}_0$. Then the wave function can be explicitly written down and the tunneling splitting expression can be derived. This is called LSLA, locally separable linear approximation [30], in which the Gamov factor is calculated first along the caustic line from the hyperbolic umbilic point to $\mathbf{x}_0^L$ (left-side taking-off point), then along the classical trajectory from $\mathbf{x}_0^L$ to $\mathbf{x}_0^R$ (right-side taking-off point), and finally along the caustic line from $\mathbf{x}_0^R$ to the right-side hyperbolic umbilic point.

In the case of Figures 3.2(b) and (c), the classically forbidden C-regions play essential roles. In the C_1-region, for instance, the motion in the η direction is nonclassical, while the motion in the ξ direction is still classical. That is why the tunneling in C-region is called *mixed tunneling*. On the other hand, in the classically forbidden I-region the motions in both directions are nonclassical and thus the tunneling in this region is called *pure tunneling*. This is the most remarkable characteristic of multidimensional tunneling. If the coupling γ becomes small, the tunneling picture changes to one like that in Figure 3.2(b). Tunneling proceeds mainly in the C_2-region and the tunneling path cannot be defined. Figure 3.2(c) shows schematically the case of large ω_y, in which tunneling mainly goes through the C_1-region. In the context of transition state theory [10,11], there have been proposed some methods to treat tunneling in chemical reaction dynamics. Here we pick up two representative ones for comparison: the Marcus-Coltrin method [63] and the Ovchinikova method [64–67]. The Marcus-Coltrin method is based on the adiabatic approximation, in which tunneling motion along the intrinsic reaction coordinate (IRC) is assumed to be much slower than the perpendicular vibrational motion and tunneling is assumed to occur along the effective one-dimensional potential, that is, the bare potential along the IRC plus the vibrational ground-state energy. This gives a reasonable description only when the curvature of IRC is small. Noting that the adiabatic approximation is all right for a large vibrational frequency, we can easily guess that Figure 3.2(c) corresponds to the situation where the Marcus-Coltrin method works. On the other hand, the Ovchinikova method may be reasonable in the opposite limit. Namely, this is based on the sudden approximation and corresponds to a large curvature case. The similar situation occurs in the case of Figure 3.2(b). Peculiarities of multidimensional

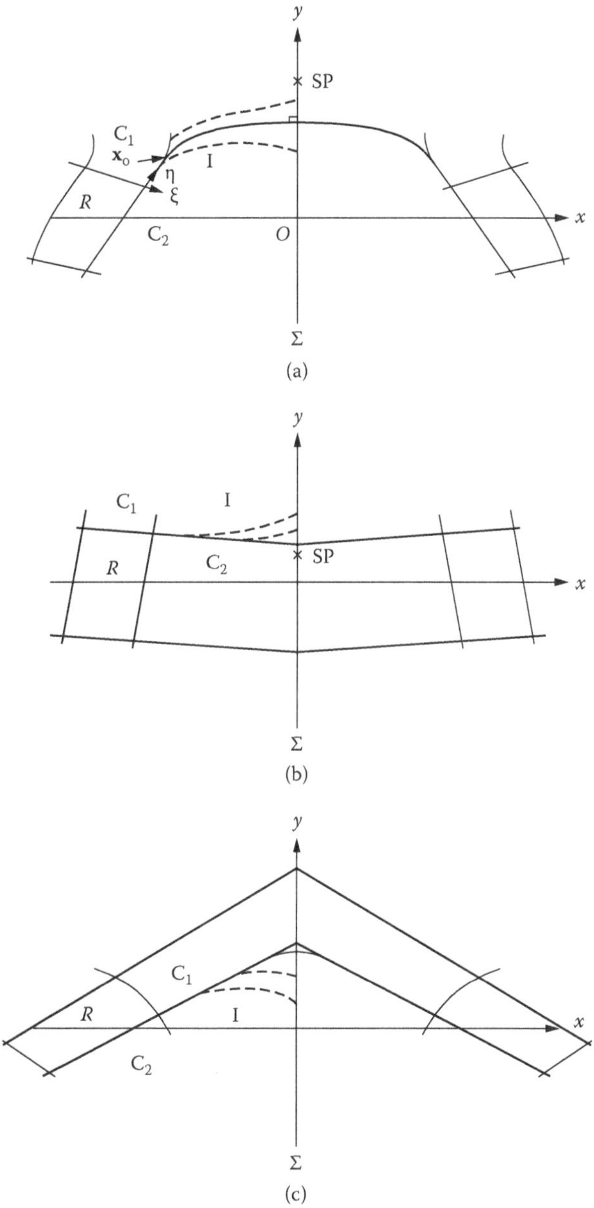

FIGURE 3.2 Schematic picture of tunneling in the case of symmetric mode coupling [see Equation (3.24)]. SP is the saddle point and Σ is the symmetry line. Dashed lines are solutions of the Hamilton-Jacobi equations but not the optimal tunneling path. (a) Case of modest ω_y and γ, in which the pure tunneling I-region plays an essential role. $\mathbf{x}_0$ is the point for the tunneling path (solid line) to depart from the caustics. (b) Case of small coupling γ, in which the mixed tunneling C_2-region plays a key role. Tunneling path cannot be defined. (c) Case of large ω_y, in which the mixed tunneling C_1-region plays a key role. (Taken from Reference [30] with permission.)

tunneling are discussed in more detail in Chapter 4, including the adiabatic and sudden approximations mainly with use of numerical calculations.

In concluding this section, it should be mentioned that the theory presented here is a general WKB theory and can, in principle, be extended to general N-dimensional ($N \geq 3$) systems. Practical applications to multidimensional systems are, however, unfortunately, quite difficult. This is mainly because there appear quite complicated features of combinations of I- and C-regions, and the propagation of the wave functions in the mixed tunneling C-region, namely, solving the nonlinear Hamilton-Jacobi equations, Equations (3.13) and (3.14), becomes a formidable task.

3.2 INSTANTON THEORY

Generalization of the one-dimensional instanton theory to two and higher dimensions has been naturally tried by many authors [17,40–42]. Here the two-dimensional extension is briefly explained according to the work by Benderskii et al. [17]. First, let us consider the decay of a metastable state. The direct two-dimensional (Q_1, Q_2) generalization of Equations (2.142) and (2.143) is given by

$$k = B\exp[-S_0/\hbar] \tag{3.25}$$

and

$$B = \sqrt{\frac{S_0}{2\pi}}\left[\frac{\det'\left(-\frac{d^2}{d\tau^2}\mathbf{1} + \mathbf{V}''(\tau)\right)}{\det\left(-\frac{d^2}{d\tau^2}\mathbf{1} + \mathbf{K}_0\right)}\right]^{-1/2}, \tag{3.26}$$

where $\mathbf{1}$ is the 2×2 unit matrix and $\mathbf{K}_0$ is a matrix of the form

$$\mathbf{K}_0 = \begin{pmatrix} \omega_+^2 & 0 \\ 0 & \omega_-^2 \end{pmatrix}. \tag{3.27}$$

This matrix is simply the diagonal form of the matrix $\partial^2 V/\partial Q_i \partial Q_j$ taken at the minimum of the potential well $\mathbf{Q} = 0$ and $\omega_\pm$ correspond to the normal coordinates (Q_+, Q_-) there with $\omega_+^2 > \omega_-^2$. Since Q_+ goes to zero faster than Q_-, asymptotic instanton direction coincides with Q_-. The important process to treat Equation (3.26) is to divide the space into transverse and longitudinal parts and deal with each of them separately. Introducing the coordinate s along the instanton trajectory and x, which measures the deviation from the instanton, and assuming that the transverse motion around the instanton trajectory is harmonic, the action S is expanded to the second order as

$$\begin{aligned} S \simeq S_0 &+ \frac{1}{2}\int_{-\infty}^{\infty} d\tau\delta s(\tau)\left(-\frac{\partial^2}{\partial\tau^2} + \frac{\partial^2 V}{\partial s^2}\right)\delta s(\tau) \\ &+ \frac{1}{2}\int_{-\infty}^{\infty} d\tau\delta x(\tau)\left(-\frac{\partial^2}{\partial\tau^2} + \omega_t^2[s(\tau)]\right)\delta x(\tau), \end{aligned} \tag{3.28}$$

where ω_t is an s-dependent transverse vibrational frequency. Then the rate constant k becomes

$$k = B_t k_{1D}, \tag{3.29}$$

where k_{1D} is the one-dimensional rate constant obtained in Section 2.5 and the transverse prefactor is given by

$$B_t = \left[\frac{\det\left(-\frac{\partial^2}{\partial \tau^2} + \omega_t^2\right)}{\det\left(-\frac{\partial^2}{\partial \tau^2} + \omega_+^2\right)} \right]^{-1/2}. \tag{3.30}$$

The zero mode discussed in Section 2.5.1 is associated with the longitudinal part and B_t does not suffer from that. The definitions of the determinants in the above equation are the same as before [see Equation (2.91)] and their estimation can be done also in the same way as before. It should be noted that if the range of the time τ is taken to be finite $(-\beta/2, \beta/2)$, then the above expressions provide the rate constant at a given temperature $T = 1/(\kappa\beta)$, where κ is the Boltzmann constant [17]. The tunneling splitting in a two-dimensional symmetric double well potential can be derived in the same way as above [17].

The extensions of the above two-dimensional theories and applications to large molecules have been carried out by many authors [40–42]. Although these may be good for qualitative understanding of multidimensional tunneling phenomena, any one of them cannot be considered to be satisfactory, unfortunately. As was mentioned in Chapter 1, the difficulties consist of (1) accurate determination of multidimensional instanton trajectory, (2) proper evaluation of the pre-exponential factor B by taking into account the effects of multidimensional transverse modes, and (3) accurate information of multidimensional potential energy surface. In the above treatments, the following basic assumptions are tacitly made: Namely, contributions of the transverse modes are additive and multidimensional systems can be treated as a set of two-dimensional ones. The fact that these are actually not true is demonstrated in Chapter 6. Besides, very accurate quantum chemical calculations are required to determine potential energy surfaces if we want to directly compare them with experimental data of polyatomic molecules. These points are discussed in Chapter 6.

4 Multidimensional Effects: Peculiar Phenomena

4.1 EFFECTS OF VIBRATIONAL EXCITATION ON TUNNELING SPLITTING

As was pointed out in Section 3.1, there are some interesting features in multidimensional tunneling compared with the one-dimensional case [30,31,46]. These are the following: (1) There exist two types of tunneling, pure tunneling and mixed tunneling. (2) No tunneling path can be defined in the case of mixed tunneling. (3) Vibrational excitation can suppress tunneling because of mixed tunneling. In this section, these peculiar features are numerically demonstrated and analyzed [31] by taking tunneling splitting in symmetric double well potential as an example. Three typical two-dimensional model potentials are used.

4.1.1 ADIABATIC AND SUDDEN APPROXIMATIONS

Since adiabatic and sudden approximations are frequently used in the study of multidimensional tunneling [4,63–69], these are briefly explained before introducing the three models.

Suppose we have a symmetric double well potential $V(x, y)$ with $V(-x, y) = V(x, y)$ (see Figure 2.9). The Hamiltonian is given by

$$H(x, y) = -\frac{\hbar^2}{2m}\left(\frac{\partial^2}{\partial x^2} + \frac{\partial^2}{\partial y^2}\right) + V(x, y). \tag{4.1}$$

Let us first consider the adiabatic approximation. The frequency ω_x in the x direction near the potential minimum is assumed to be much smaller than the frequency ω_y in the y direction. Then the x motion is frozen and the adiabatic solutions $\chi_{n_y}(y; x)$ that satisfy the following equation are solved first:

$$\left[-\frac{\hbar^2}{2m}\frac{\partial^2}{\partial y^2} + V(x, y)\right]\chi_{n_y}(y; x) = \epsilon_{n_y}(x)\chi_{n_y}(y; x). \tag{4.2}$$

The total wave function is assumed to be given by

$$\Psi_{\pm,n_x,n_y}(x, y) = \phi_{\pm,n_x n_y}(x)\chi_{n_y}(y; x), \tag{4.3}$$

where $\pm$ specifies the parity with respect to the y axis. Inserting this into the Schrödinger equation and neglecting the derivative of $\chi_{n_y}(y; x)$ with respect to x, we obtain

$$\left[-\frac{\hbar^2}{2m}\frac{\partial^2}{\partial x^2} + \epsilon_{n_y}(x)\right]\phi_{\pm,n_x n_y}(x) = E_{\pm,n_x n_y}\phi_{\pm,n_x n_y}(x). \tag{4.4}$$

The tunneling splitting can be obtained as

$$\Delta E_{n_x n_y} = \Delta E_{n_x}^{1D}[\epsilon_{n_y}(x)], \tag{4.5}$$

where the right-hand side (RHS) is the energy splitting in the one-dimensional potential $\epsilon_{n_y}(x)$.

Let us next consider the sudden approximation. The Hamiltonian is the same as Equation (4.1), but now the frequency ω_y is assumed to be much smaller than ω_x. The total wave function can be written as

$$\Psi_{\pm,n_x n_y}(x, y) = \phi_{\pm,n_x n_y}(y)\chi_{\pm,n_x}(x; y), \tag{4.6}$$

where the adiabatic bases $\chi_{\pm,n_x}(x; y)$ and the coefficient function $\phi_{\pm,n_x n_y}(y)$ are defined in the analogous way as Equations (4.2) and (4.4) by

$$\left[-\frac{\hbar^2}{2m}\frac{\partial^2}{\partial x^2} + V(x, y)\right] \chi_{\pm,n_x}(x; y) = \epsilon_{\pm,n_x}(y)\chi_{\pm,n_x}(x; y) \tag{4.7}$$

and

$$\left[-\frac{\hbar^2}{2m}\frac{\partial^2}{\partial y^2} + \epsilon_{\pm,n_x}(y)\right] \phi_{\pm,n_x n_y}(y) = E_{\pm,n_x n_y}\phi_{\pm,n_x n_y}(y). \tag{4.8}$$

Since ϕ is not directly related to tunneling, we approximate Equation (4.8) by the equation

$$\left[-\frac{\hbar^2}{2m}\frac{\partial^2}{\partial y^2} + V(x_{min}, y)\right] \phi_{n_y}(y) = E'_{n_y}\phi_{n_y}(y), \tag{4.9}$$

where x_{min} is the x coordinate of the potential minimum. Taking symmetric and antisymmetric linear combinations of $\Psi_{\pm,n_x n_y}$, we can obtain the wave functions localized in each well as

$$\Psi^L_{n_x n_y}(x, y) = \phi_{n_y}(y)\chi^L_{n_x}(x; y) \tag{4.10}$$

and

$$\Psi^R_{n_x n_y}(x, y) = \phi_{n_y}(y)\chi^R_{n_x}(x; y). \tag{4.11}$$

Inserting these into the Herring formula, we obtain the final expression of tunneling splitting,

$$\Delta E_{n_x n_y} = \int dy \phi^2_{n_y}(y)\Delta E^{1D}_{n_x}(y). \tag{4.12}$$

This is similar to the Franck-Condon approximation. Tunneling in the x direction occurs at each fixed value of y, which creates the tunneling splitting $\Delta E^{1D}_{n_x}(y)$. The total splitting is obtained by averaging $\Delta E^{1D}_{n_x}(y)$ over y with the weight factor $\phi^2_{n_y}(y)$.

4.1.2 Case of Symmetric Mode Coupling Potential

Let us next consider the symmetric mode coupling (SMC) potential. The Hamiltonian is given by

$$\tilde{H} = -\frac{\hbar^2}{2m_x}\frac{\partial^2}{\partial \tilde{x}^2} - \frac{\hbar^2}{2m_y}\frac{\partial^2}{\partial \tilde{y}^2} + \frac{m_x\tilde{\omega}_x^2}{8\tilde{x}_0^2}(\tilde{x} - \tilde{x}_0)^2(\tilde{x} + \tilde{x}_0)^2 + \frac{1}{2}m_y\tilde{\omega}_y^2[\tilde{y} + \tilde{\alpha}(\tilde{x}^2 - \tilde{x}_0^2)]^2, \tag{4.13}$$

where $m_x, m_y, \tilde{\omega}_x, \tilde{\omega}_y, \tilde{x}_0$, and $\tilde{\alpha}$ are mass in x direction, mass in y direction, frequency in x direction, frequency in y direction, one-half of the distance between two minima, and coupling strength, respectively. Introducing the following dimensionless quantities, the Schrödinger equation can be scaled:

$$x = \frac{\tilde{x}}{\tilde{x}_0}, \quad y = \sqrt{\frac{m_x}{m_y}}\frac{\tilde{y}}{\tilde{x}_0},$$

$$\omega_y = \frac{m_y \tilde{\omega}_y}{m_x \tilde{\omega}_x}, \quad g = \frac{\hbar}{m_y \tilde{\omega}_x \tilde{x}_0^2},$$

$$\alpha = \tilde{\alpha}\tilde{x}_0\sqrt{\frac{m_x}{m_y}}, \quad H = \frac{\tilde{H}}{m_x \tilde{\omega}_x^2 \tilde{x}_0^2}. \tag{4.14}$$

Then the scaled Hamiltonian with the symmetric mode coupling (SMC) potential is given by

$$H_{SMC} = -\frac{g^2}{2}\left(\frac{\partial^2}{\partial x^2} + \frac{\partial^2}{\partial y^2}\right) + \frac{1}{8}(x-1)^2(x+1)^2 + \frac{\omega_y^2}{2}\left(y + \alpha(x^2-1)\right)^2, \tag{4.15}$$

where g plays the role of $\hbar$. This Hamiltonian is the same as Equation (3.24). The potential has two minima $(x, y) = (\pm 1, 0)$ and one saddle point at $(x = 0, y = \alpha)$. The barrier height is normalized to 1/8. The frequency in y direction is always $(\partial^2 V_{SMC}/\partial y^2)^{1/2} = \omega_y$, while that in x direction at the minima is normalized to unity. The potential contour is depicted in Figure 4.1 for $(\omega_y, \alpha) = (0.5, 1.0)$. This model has been used as a typical example of multidimensional tunneling in many fields [4,56]. Especially, in the case of proton tunneling in malonaldehyde, the coordinate x mainly represents the motion of the hydrogen atom transferring from one oxygen atom O to the other oxygen atom O, while y roughly represents the scissors-like motion of the O–C–C–C–O frame [see Figure 4.2(a)]. The fact that the mutual approach of the two oxygen atoms makes the tunneling more probable corresponds to tunneling enhancement by excitation in y mode. More generally, the a_1 symmetric mode of a nonrigid molecule coupled to a tunneling coordinate assists the hydrogen tunneling [70]. Table 4.1 shows the results of quantum mechanically exact numerical calculations of tunneling splitting for two sets of parameters: (1) the intermediate coupling case $[(\omega_y, \alpha) = (0.8, 1.0)]$ in which the tunneling mainly goes through the I-region [see Figure 3.2(a)] and (2) the weak coupling case $[(\omega_y, \alpha) = (0.3, 0.22222)]$ in which tunneling mainly goes through the C-region [see Figure 3.2(b)]. In the first case, vibrational excitation promotes the tunneling very much. This can be simply attributed to the decrease of the Gamov factor. This is a common phenomenon that is well known in the one-dimensional case. The second case also shows the promotion of tunneling by the vibrational excitation. However, the promotion is weak. This is simply because the coupling is weak. Actually, no qualitative difference between the two cases is found in this symmetric mode coupling potential.

It is of great interest to see how the tunneling path is affected by vibrational excitation when the tunneling goes through the I-region. Figure 4.3 shows this for several vibrational states in the case of the V-parabola model with parameters $(\omega_s, \theta, g) = (0.42, 30°, 0.04)$. The V-parabola model is simpler than the symmetric mode coupling

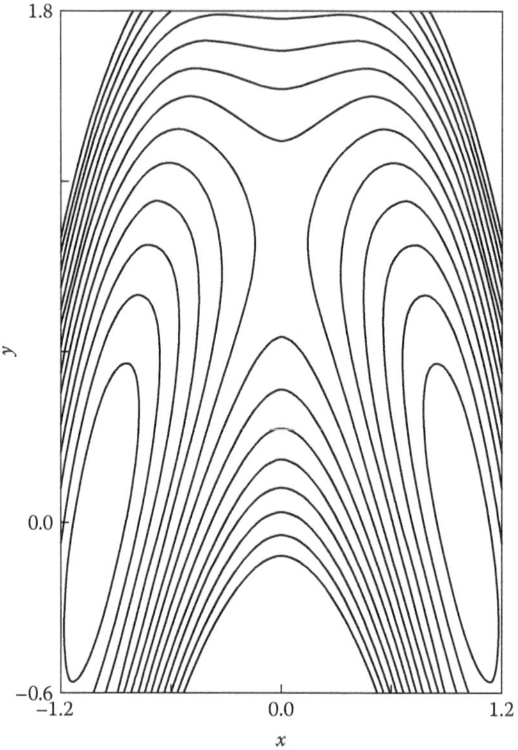

FIGURE 4.1 Contours of the symmetric mode coupling potential [Equation (4.15)] for the parameters $(\omega_y, \alpha) = (0.5, 1.0)$. The interval of contour is 0.02 and the contours with the energy higher than 0.28 are omitted. (Taken from Reference [31] with permission.)

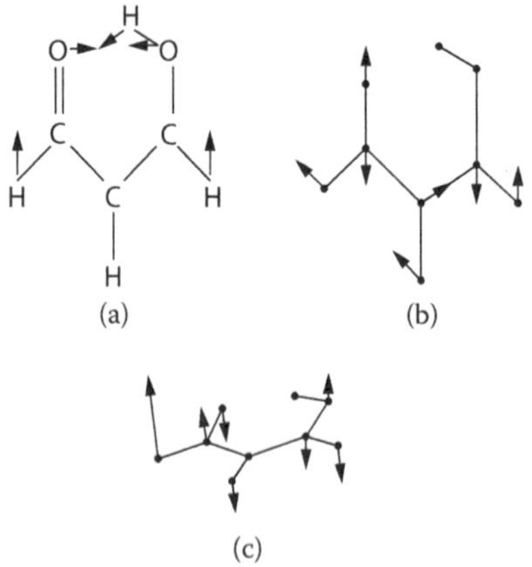

FIGURE 4.2 Various modes coupled to the hydrogen tunneling mode in the case of malonaldehyde. (a) In-plane scissors-like mode corresponding to the symmetric mode coupling model. (b) In-plane antisymmetric C–O stretching mode corresponding to the asymmetric mode coupling model. (c) Out-of-plane wagging motion corresponding to the squeezed potential model. (Taken from Reference [31] with permission.)

TABLE 4.1
Tunnel Splitting in the Symmetric Mode Coupling Potential Equation (4.15) Calculated Quantum Mechanically Numerically for Several Vibrationally Excited States. Parameter g Is Taken to be 0.04

(ω_y, α)	**(0.8, 1.0)**	**(0.3, 0.22222)**
$\Delta E_{0,0}$[a]	4.6(−9)[b]	2.33(−8)
$\Delta E_{0,1}$	122.8(−9)	3.07(−8)
$\Delta E_{0,2}$	167.1(−9)	3.93(−8)
$\Delta E_{0,3}$	15301.1(−9)	4.93(−8)
Tunneling regions	*I–I*	*C–C*

[a] $\Delta E_{0,n}$ indicates the tunneling splitting for the state in which the vibrational motions are in the ground (0) and the nth excited levels for the tunneling direction and the direction transversal to it, respectively.
[b] Figures in parentheses are the power of 10 by which the entry is to be multiplied.

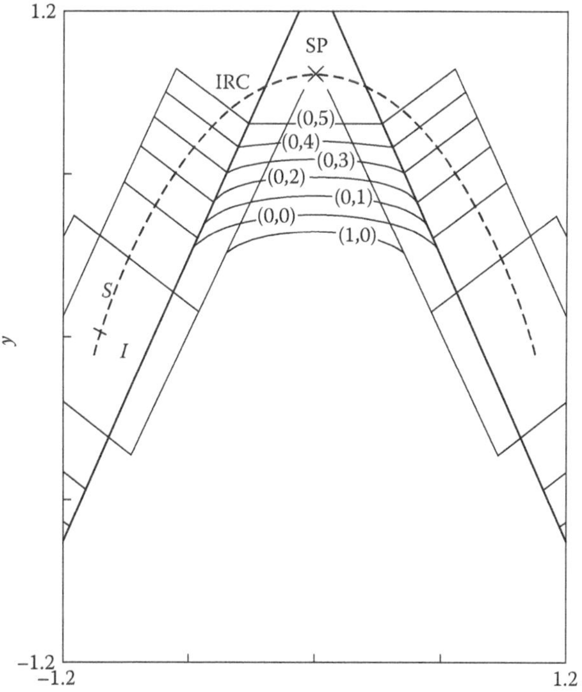

FIGURE 4.3 Tunneling paths for several vibrationally excited states in the case of the V-parabola model [Equation (4.16)]. Parameters are chosen to be $(\omega_s, \theta, g) = (0.42, 30°, 0.04)$. (n_l, n_s) specifies the vibrational quantum numbers of l and s modes. (Taken from Reference [31] with permission.)

potential but has similar characteristics. The potential is parabolic in both left ($x < 0$) and right ($x > 0$) sides, and is given by

$$V_{VP}(x, y) = \frac{1}{2}\omega_s^2 + \frac{1}{2}l^2, \tag{4.16}$$

where the kinetic energy part is the same as Equation (4.15) and the coordinates (s, l) are related to (x, y) as

$$\begin{pmatrix} s \\ l \end{pmatrix} = \begin{pmatrix} \cos\theta & -\sin\theta \\ \sin\theta & \cos\theta \end{pmatrix} \begin{pmatrix} x \pm 1 \\ y \end{pmatrix}. \tag{4.17}$$

Here the sign + (−) is taken in the left- (right-)hand side. The parameter θ roughly plays the same role as the coupling strength α in the symmetric mode coupling model. This potential has a cusp along the y axis, but the essential conclusions are not affected by this. As is seen from Figure 4.3, the tunneling paths are strongly dependent on vibrational state. The number in the parenthesis (n_s, n_l) on each path represents the vibrational quantum numbers in the l and s modes, respectively. All the tunneling paths take shortcuts compared with the intrinsic reaction coordinate (IRC). The two vibrational modes have opposite effects: excitation in mode s makes tunneling path closer to the IRC, while that in l mode shifts it inward.

Finally, let us investigate the validity of the adiabatic and sudden approximations. If we apply the adiabatic approximation to the symmetric mode coupling potential, the tunneling splitting becomes independent of the vibrational quantum number n_y, whatever ω_y is. The adiabatic potential has the same shape as that of $n_y = 0$, since the frequency in y direction is constant. This is not in agreement with the exact quantum results except when the coupling α is zero. Table 4.2 compares the tunneling splittings calculated by the sudden approximation with the exact results. In the sudden approximation, $\Delta E_{n_y}^{1D}(y)$ in Equation (4.12) is calculated by the WKB formula. It is seen that tunneling splitting by sudden approximation decreases with vibrational excitation, while the exact one increases. Thus, neither the adiabatic nor sudden approximation

TABLE 4.2
Tunnel Splitting in the Symmetric Mode Coupling Potential Equation (4.15). The Second (Third) Column Is the Result by Exact Quantum Mechanical (EQM) Calculation [by the Sudden Approximation Equation (4.12)]. Parameters (ω_y, α, g) Are Equal to (0.2,0.25,0.04)

	EQM	Sudden
$\Delta E_{0,0}$	2.36(−8)	2.18(−8)
$\Delta E_{0,1}$	2.60(−8)	1.88(−8)
$\Delta E_{0,2}$	2.85(−8)	1.26(−8)
$\Delta E_{0,3}$	3.10(−8)	1.00(−8)

is reliable for investigating the tunneling in excited states of the symmetric mode coupling (SMC) model.

4.1.3 Case of Antisymmetric Mode Coupling Potential

The scaled Hamiltonian with the antisymmetric mode coupling (ASMC) potential is defined by

$$H_{ASMC} = -\frac{g^2}{2}\left(\frac{\partial^2}{\partial x^2} + \frac{\partial^2}{\partial y^2}\right) + \frac{1}{8}(x-1)^2(x+1)^2 + \frac{\omega_y^2}{2}(y - \beta x)^2, \quad (4.18)$$

where g and ω_y have the same meanings as those of the symmetric mode coupling (SMC) model, Equation (4.15), and β represents the coupling strength. A potential contour map is given in Figure 4.4. There are two minima at $(x, y) = (\pm 1, \pm\beta)$ and one saddle point at $(x, y) = (0, 0)$. These two minima are shifted in y direction and the barrier height is again normalized to 1/8. In the case of malonaldehyde, the coordinate x again represents the motion of transferring hydrogen, while y represents the antisymmetric C–O stretching mode [see Figure 4.2(b)]. With use of this model, the tunneling splittings are calculated quantum mechanically. Two distinct cases with different characteristics of tunneling are shown in Table 4.3. The second column

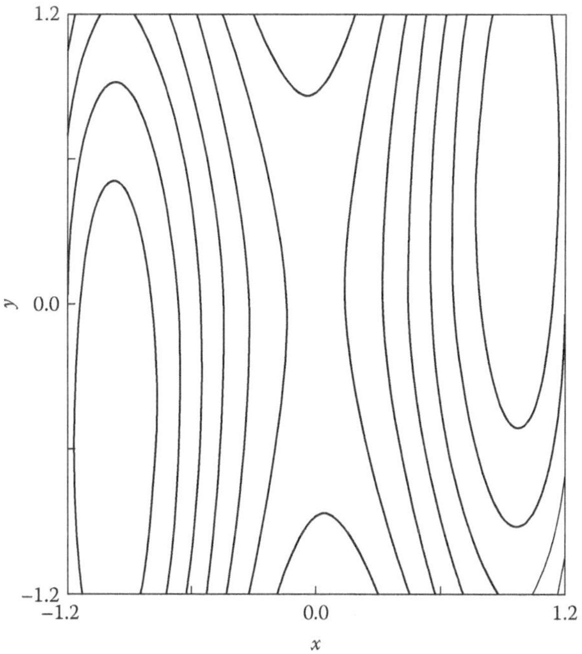

FIGURE 4.4 Contours of the antisymmetric mode coupling (ASMC) potential [Equation (4.18)] for the parameters $(\omega_y, \beta) = (0.2, 0.5)$. The other conditions are the same as in Figure 4.1. (Taken from Reference [30] with permission.)

TABLE 4.3
Tunnel Splitting in the Antisymmetric Mode Coupling Potential Equation (4.18) Calculated Quantum Mechanically Numerically. Parameter g Is Equal to 0.04

(ω_y, β)	**(0.5, 1.0)**	**(0.2, 0.3)**
$\Delta E_{0,0}$	1.4(−10)	182.1(−10)
$\Delta E_{0,1}$	60.5(−10)	5.7(−10)
$\Delta E_{0,2}$	1316.5(−10)	85.0(−10)
$\Delta E_{0,3}$	18920.0(−10)	117.9(−10)
$\Delta E_{0,4}$		113.9(−10)
$\Delta E_{0,5}$		88.7(−10)
Tunneling regions	*I–I*	*C–C*

represents the case in which tunneling mainly goes through the I-region while the third column corresponds to the interesting case in which tunneling mainly passes through the C-region. In the case of pure tunneling through the I-region, tunneling is always promoted by vibrational excitation and that is attributed to the decrease of the Gamov factor. In the case of mixed tunneling through the C-region, on the other hand, the tunneling splitting oscillates with respect to n_y and is sensitive to the parameter β. This is definitely different from the case of symmetric mode coupling (SMC) model. This oscillatory feature occurs because the excited vibrational wave functions have nodal lines in the C-region (nearly parallel to the x axis) and thus the overlap integral in the Herring formula can oscillate due to the phase cancellation.

Finally, we touch upon the adiabatic and sudden approximations in the present model. In the same way as in the case of symmetric mode coupling model, the adiabatic approximation leads to the tunneling splitting independent of n_y. This does not exhibit any characteristic behavior discussed above and can never be reliable. If we apply the sudden approximation, we also encounter a problem since the potential curve in x direction is not symmetric except when y is zero. Thus we cannot use Equation (4.12) directly anymore.

4.1.4 Case of Squeezed (Sqz) Double Well Potential

As a third example, we consider the squeezed (Sqz) model potential whose scaled Hamiltonian is given by

$$H_{SQZ} = -\frac{g^2}{2}\left(\frac{\partial^2}{\partial x^2} + \frac{\partial^2}{\partial y^2}\right) + \frac{1}{8}(x-1)^2(x+1)^2 + \frac{1}{2}[\omega_y^2 - \gamma(x^2-1)]y^2, \quad (4.19)$$

where g and ω_y have the same meanings as those of symmetric mode coupling model [see Equation (4.15)] and γ represents the coupling strength. Two minima at $(x, y) = (\pm 1, 0)$ and one saddle point at $(x, y) = (0, 0)$ are located on the x axis. The barrier height is again normalized to 1/8. The local frequency along the y axis is

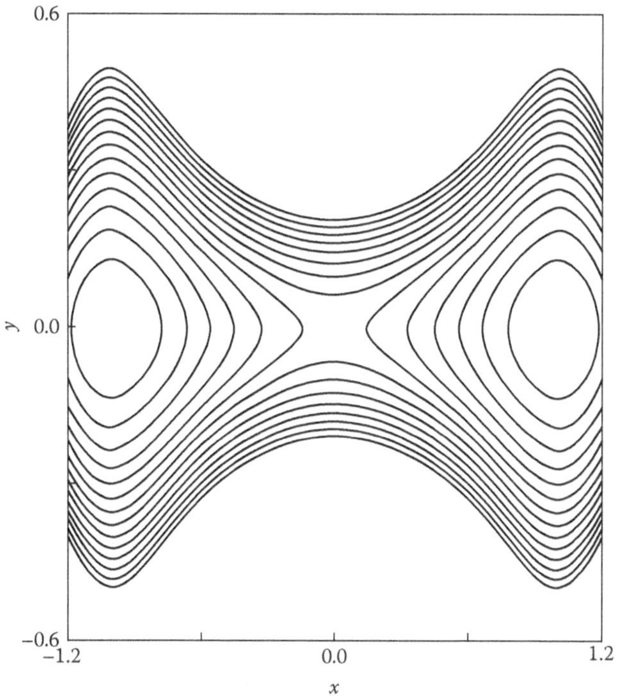

FIGURE 4.5 Contours of the squeezed (Sqz) potential [Equation (4.19)] for the parameters $(\omega_y, \gamma) = (1.5, 1.0)$. The other conditions are the same as in Figure 4.1. (Taken from Reference [30] with permission.)

defined as

$$\Omega_y^2(x) = \omega_y^2 - \gamma(x^2 - 1), \tag{4.20}$$

and thus the coupling parameter γ plays a role of squeezing (positive γ) and spreading (negative γ) the potential in y direction. Figure 4.5 depicts the potential contour for the case $(\omega_y, \gamma) = (1.5, 1.0)$. It should be noted that when $\omega_y^2 > -\gamma$, this potential has additional saddle points at $(x, y) = (\pm\sqrt{1 + \omega_y^2/\gamma}, \pm\omega_y/\sqrt{2}|\gamma|)$ and drops beyond these points. Thus we cannot take a very large positive γ so as to keep the double well topography. A typical example of the squeezed potential can be found also in malonaldehyde, in which the y coordinate now corresponds to an out-of-plane wagging motion depicted schematically in Figure 4.2(c). Since the stable structure of malonaldehyde is planar and the out-of-plane deviation makes the O–O distance larger, the topography of the potential energy surface should be represented by this squeezed (Sqz) mode potential. Exact quantum mechanical calculations have been carried out and the results for a representative parameter are shown in Table 4.4. The following interesting features can be found: (1) vibrational excitation in the y direction suppresses tunneling when the vibrational quantum number n_y is small ($n_y \leq 4$ in the present case), (2) the amount of suppression by excitation decreases with n_y, and (3) the splitting starts to increase from a certain n_y ($n_y = 5$ in the present case). In

TABLE 4.4
Tunnel Splitting in the Squeezed Potential Equation (4.19) Calculated with Use of the Exact Quantum Mechanical (EQM) Method and the Sudden Approximation (Sudden). Parameters (ω_y, γ, g) Are Equal to (0.2, 0.05, 0.05)

Method	EQM	Sudden
$\Delta E_{0,0}$	6.12(−7)	6.01(−7)
$\Delta E_{0,1}$	3.99(−7)	4.08(−7)
$\Delta E_{0,2}$	3.31(−7)	3.07(−7)
$\Delta E_{0,3}$	3.04(−7)	2.49(−7)
$\Delta E_{0,4}$	3.03(−7)	2.14(−7)
$\Delta E_{0,5}$	3.09(−7)	1.90(−7)

the sudden regime $\omega_y \ll 1$, feature (1) can be understood intuitively based on the sudden approximation [see Equation (4.12)]. As n_y increases, the wave function $\phi_{n_y}(y)$ spreads and the population near the x axis decreases. On the other hand, $\Delta E^{1D}(y)$ decreases rapidly as $|y|$ increases, since the potential is squeezed in y direction. Thus, the Franck-Condon integral is expected to decrease as n_y increases. Actually, the tunneling splitting by the sudden approximation given in the third column of Table 4.4 shows the suppression of tunneling by vibrational excitation in the y direction.

When ω_y is larger than unity, the adiabatic regime holds. Table 4.5 shows some interesting cases. Tunneling splittings are given for a small γ (second and third columns) and for a relatively large γ (fourth column). In the case of small coupling γ, the tunneling is suppressed by the vibrational excitation. This can be interpreted by the adiabatic approximation, Equation (4.5), namely, in terms of the renormalized potential for each n_y shown in Figure 4.6. From this figure we can predict that the tunneling splitting is diminished by vibrational excitation in the y direction. Actually, the tunneling splittings estimated by Equation (4.5) (third column of

TABLE 4.5
Tunneling Splitting in the Squeezed (Sqz) Potential Equation (4.19) with Large ω_y. The Parameter g Is Taken to Be 0.05

(ω_y, γ)	(1.8, 0.1)		(1.5, 0.5)
Method	EQM	Adiabatic	EQM
$\Delta E_{0,0}$	7.16(−7)	6.92(−7)	5.57(−7)
$\Delta E_{0,1}$	6.45(−7)	6.24(−7)	3.01(−7)
$\Delta E_{0,2}$	5.84(−7)	5.63(−7)	83.7(−7)

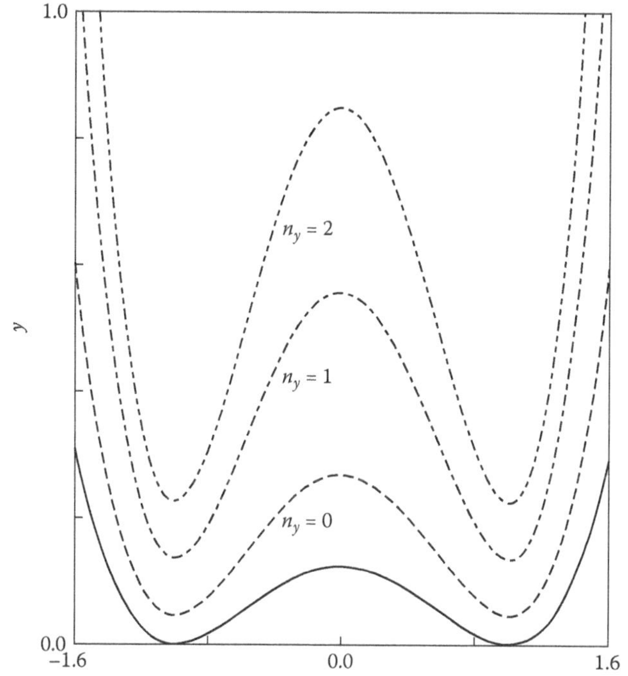

FIGURE 4.6 Renormalized one-dimensional potential curve [see Equation (4.21)]. Dashed lines represent $V_{n_y}^{ren}(x)$, while the solid line is the bare potential $V_{Sqz}(x, 0)$. (Taken from Reference [30] with permission.)

Table 4.5) agree quite well with the exact quantum mechanical (EQM) results. The renormalized potential is defined as follows:

$$V_{n_y}^{ren} = V_{SQZ}(x, 0) + g\Omega_y(x)\left(n_y + \frac{1}{2}\right), \tag{4.21}$$

where the local frequency $\Omega_y(x)$ is defined by Equation (4.20).

If we use the one-dimensional WKB theory, ΔE_0 is obtained as

$$\Delta E_0 \simeq \frac{g}{\pi} \exp\left(-\frac{1}{g}\int_{x_-}^{x_+} dx \sqrt{2[V_0^{ren}(x) - E]}\right), \tag{4.22}$$

where $x_\pm$ are solutions of $V_0^{ren}(x) - E = 0$. It should be noted that the effect of potential squeezing is found in the exponent.

As the coupling γ increases (fourth column), this suppression of tunneling disappears and irregular behavior comes out. Another interesting feature appears when the Fermi resonance occurs between the two states $(0, n_y)$ and (n'_x, n'_y) with nonzero n'_x. Tunneling splitting itself is found to disappear in this case. Since the state with higher n_x has larger amplitude in the potential barrier region, an admixture of the state (n'_x, n'_y) in the state $(0, n_y)$ crucially affects the tunneling splitting of the latter state, even if the admixture is very small.

4.2 INSUFFICIENCY OF TWO-DIMENSIONAL MODEL

As was mentioned in Section 3.2, many of the multidimension theories are based on the assumptions that the multidimensional problem can be treated by two dimensions, one along the tunneling or instanton trajectory and the other along the transversal direction perpendicular to the tunneling trajectory, and that the contributions from other transverse modes are additive. These assumptions are, unfortunately, not necessarily consistent and sufficient. The pre-exponential factor comes out as a result of fluctuations of the classical instanton trajectory and it can be proved that the harmonic fluctuations along longitudinal and transverse directions are not coupled. Thus, in the case of two dimensions the pre-exponential factor can always be found from a solution of one-dimensional dynamical problem along the direction orthogonal to the instanton. In higher dimensions the situation changes, since different transverse classical fluctuations cannot be generally uncoupled. No matter how weak this coupling is, the classical motion along the instanton is infinitely long and the energy is redistributed between all the modes. Ignoring this effect may lead to erroneous results even for the ground state and it becomes totally inappropriate in the case of mode-specific tunneling. This can be demonstrated by the behavior of the effective frequency $\theta(\tau)$ in the direction of transversal local coordinates along the instanton. This frequency $\theta(\tau)$ sometimes deviates from the adiabatic approximation $\theta_{ad}(\tau)$, which is the local adiabatic transversal frequency along the instanton and often used in the simplified version of the instanton approach. For low-energy excitation, $\theta(\tau)$ is generally close to $\theta_{ad}(\tau)$, but for high-energy excitation the adiabatic approximation $\theta_{ad}(\tau)$ significantly deviates from $\theta(\tau)$. Such a dramatic effect of vibrational excitation cannot be reproduced by the two-dimensional models. Practical examples are shown in Chapter 7 [45].

4.3 PROTON TUNNELING IN TROPOLONE

Following the discussions by Takada and Nakamura [31], we present here the theoretical analysis of tunneling dynamics in the first excited state $\tilde{A}(^1B_2)$ of the tropolone molecule (see Figure 4.7). A three-dimensional potential model is employed and quantum mechanical numerical calculations are carried out. This kind of analysis would be useful also for other systems, unless high accuracy is required.

4.3.1 AVAILABLE EXPERIMENTAL DATA

There are quite a few experimental data on tropolone. The electronically ground state $\tilde{X}(^1A_1)$ has been studied by infrared (IR) spectroscopy [71,72], Fourier transform microwave spectroscopy [73], and single vibronic level fluorescence (SVLF) spectroscopy [74–76]. IR spectra give us information about normal modes. The SVLF spectra provide us with the relation between the vibronic level in the excited state and that in the ground state. For the first electronically excited state $\tilde{A}(^1B_2)$, the UV absorption spectra [77] and the laser fluorescence excitation spectra [70,74,76,78] have been measured. The latter have also been measured for the isotope substituted tropolone [79,80]. The data obtained from the laser fluorescence excitation spectra

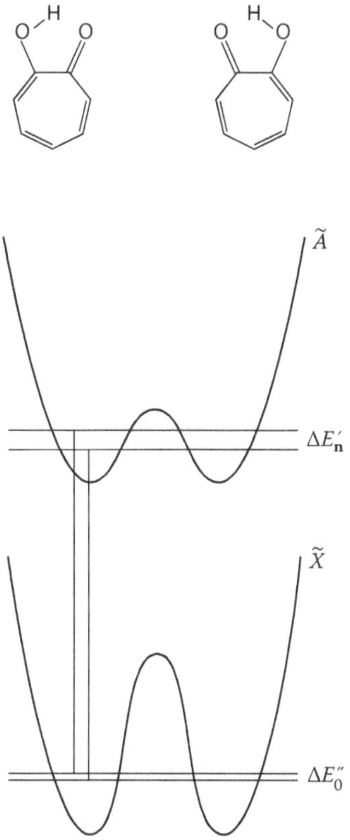

FIGURE 4.7 Tropolone molecule and the concept to determine the tunneling splitting in the excited state (see the text). (Taken from Reference [31] with permission.)

include the main information we are interested in here, namely, the accurate energy levels of various vibrationally excited states of $\tilde{A}(^1B_2)$.

The direct information of tunneling splitting in the excited state is obtained by combining the laser fluorescence excitation spectroscopy and the Fourier transform microwave spectroscopy. As is shown in Figure 4.7, the laser fluorescence excitation spectra for jet-cooled tropolone give us the values of $|\Delta E'_{\mathrm{n}} - \Delta E''_0|$, where $'$ and $''$ represent the $\tilde{A}$ and $\tilde{X}$ states, respectively. Since $\Delta E''_0$ is determined to be 0.974 cm^{-1} from the Fourier transform microwave spectra [73], we can estimate $\Delta E'_{\mathrm{n}}$. The vibrational modes of the $\tilde{A}$ state can be assigned indirectly. For the ground state $\tilde{X}$, using the results of infrared spectra and molecular orbital (MO) calculations, we can assign each vibrational mode. The correspondence between the modes of $\tilde{X}$ and $\tilde{A}$ states is established by using the single vibronic level fluorescence (SVLF) spectra. Combining these data, we can guess the characteristics of the vibrational modes in the $\tilde{A}$ state. The tunneling splittings $\Delta E'_{\mathrm{n}}$ observed and assigned in this way are listed in Table 4.6, where TRNOH (TRNOD) represents tropolone (that with $-$OH deuterated to $-$OD). Many interesting features can be found such as (1) excitation of the ν'_{12}

TABLE 4.6
Observed Tunneling Splittings $\Delta E'_n$ of Tropolone $\tilde{A}$ State

Band	TRNOH[a]	TRNOD[b]	Assignment	Symmetry[d]
0^0	20	2	Origin	
11^1	14	c	C–C stretch	a_1
12^1	18	2	CCC bend	a_1
13^1	33	3	CC–O/CCC deform	a_1
14^1	31	11	in-plane ring deform	a_1
14^2	29	13		
19^2	10	c	out-of-plane bend	a_2
25^2	5		out-of-plane bend	b_1
26^2	8	c	out-of-plane bend	b_1
26^4	6			
26^6	5			
26^8	2			
26^{10}	c			
26^{12}	c			
$14^1 26^2$	5	2		
$25^1 26^1$	c	c		

[a]References [70,73,76].
[b]Reference [76]. Here, $\Delta E''_0$ is assumed to be zero.
[c]Unresolved.
[d]Irreducible representation of C_{2V} point group. Tropolone is considered to be nonrigid. These assignments are due to Reference [72].

mode does not affect the tunneling significantly, (2) excitation of the modes ν'_{13} and ν'_{14} promote the tunneling, (3) excitation of the modes ν'_{25} and ν'_{26} suppress the tunneling, and so on. It should be noted that the zero-point tunneling splittings in the $\tilde{X}$ and $\tilde{A}$ states are also measured by Bracamonte and Vaccaro to be 0.98 cm^{-1} and 19.85 cm^{-1}, respectively [81].

4.3.2 Tunneling Dynamics in the Ground $\tilde{X}$ State

The quantum chemical MO calculations of the $\tilde{X}$ state was carried out [31] by using the GAUSSIAN 92 program package [82]. The MP2 method with 6-31G** basis set is employed to calculate the energy, the fully optimized geometries, 39 normal modes, and their frequencies at the stable structure and the saddle point. The energy is also calculated by MP4 method with 6-31G** basis set at the stable and saddle-point structures to get a more accurate estimate of the energy barrier. The details are not given here, but on the basis of these calculations, a three-dimensional model potential is constructed. It is natural to assume that the vibrational modes, which largely affect the tunneling dynamics, are basically the same for the $\tilde{X}$ and $\tilde{A}$ states. Thus, the direct tunneling mode, the mode ν''_{13}, and the mode ν''_{26} are selected as relevant coordinates

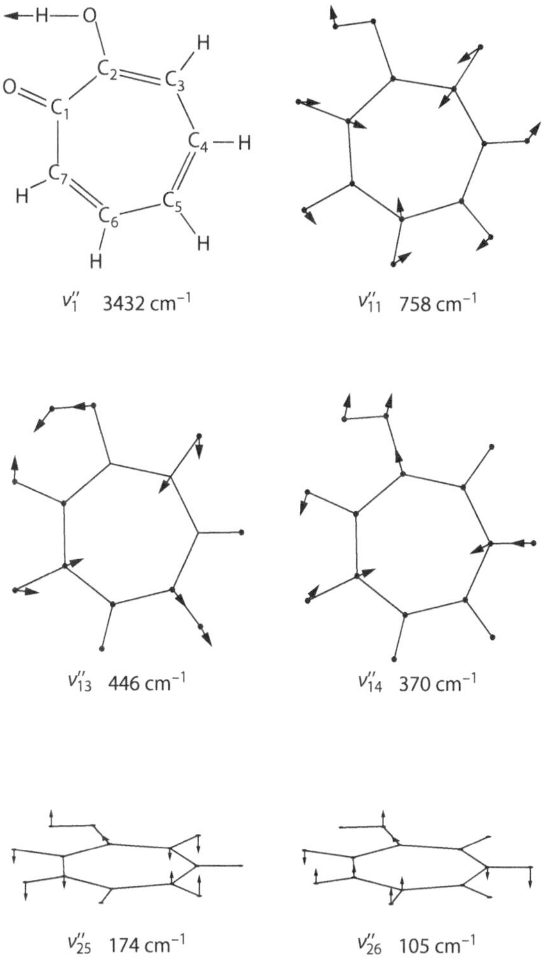

FIGURE 4.8 Normal modes of tropolone in the $\tilde{X}$ state. (Taken from Reference [31] with permission.)

for a three-dimensional model. The other modes are regarded as irrelevant ones, although their zero-point energies are added to the model potential. Normal modes in the $\tilde{X}$ state are shown in Figure 4.8. Since ν_{13}'' is assigned to a_1 symmetry, the coupling should be that of the symmetric mode coupling potential. On the other hand, ν_{26}'' is the out-of-plane wagging motion and thus the potential in this direction should be squeezed. On the basis of these considerations, the following model potential is constructed:

$$H = -\frac{g^2}{2}\left(\frac{\partial^2}{\partial x^2} + \frac{\partial^2}{\partial y^2} + \frac{\partial^2}{\partial z^2}\right) + \frac{1}{8}(x-1)^2(x+1)^2 + \frac{\omega_y^2}{2}[y + \alpha(x^2-1)]^2$$
$$+\frac{1}{2}[\omega_z^2 - \gamma(x^2-1)]z^2, \tag{4.23}$$

TABLE 4.7
Dimensionless Parameters of the Three-Dimensional Model Equation (4.23) for the Tropolone $\tilde{X}$ State

Parameter	Value
g	0.0732
ω_y	0.469
α	1.19
ω_z	0.0446
γ	0.00190

where modes x and z represent the direct tunneling mode and ν''_{26} mode, respectively. The mode y collectively represents the coupling of a_1 mode and largely includes ν''_{13} mode.

The dimensionless parameters of the above Hamiltonian are determined from the quantum chemical data obtained by removing the translational and rotational degrees of freedom [31]. These parameters are listed in Table 4.7. No parameter was fitted to reproduce any tunneling splitting data. Tunneling splittings are calculated quantum mechanically numerically with this three-dimensional Hamiltonian using the discrete variable representation (DVR) method [83]. Numbers of grids used here are 60, 40,

TABLE 4.8
Tunneling Splitting (cm^{-1}) Calculated with Use of the Exact Quantum Mechanical (EQM) Numerical Method for the Model Potential Equation (4.23) of Tropolone

Splitting	EQM
$\Delta E_{0,0,0}$[a]	1.05
$\Delta E_{0,0,1}$	0.965
$\Delta E_{0,0,2}$	0.887
$\Delta E_{0,0,3}$	0.819
$\Delta E_{0,0,4}$	0.758
$\Delta E_{0,0,5}$	0.705
$\Delta E_{0,0,6}$	0.656
$\Delta E_{0,1,0}$	7.61
$\Delta E_{0,1,1}$	7.02

[a] Subscripts represent vibrational quantum numbers of the direct tunneling mode, ν''_{13} mode (approximately), and ν''_{26} mode.

and 40 for x, y, and z coordinates, respectively. Table 4.8 gives the tunneling splittings for several vibrational states. First, the tunneling splitting of the ground state ΔE_0 is 1.05 cm^{-1}, which is in good agreement with observed values (= 0.974 cm^{-1} [73], 0.98 cm^{-1} [81]). It may be fair to say that this good agreement is somewhat accidental, because the present model is simple and tunneling splitting is known to be very sensitive to the accuracy of potential energy surface. The more important thing is the dependence of $\Delta E_{\mathbf{n}}$ on $\mathbf{n}$. It can be seen that the vibrational excitation in z direction suppresses the tunneling weakly, while that in y direction promotes it quite strongly. The modes y and z play the same roles as those in the two-dimensional symmetric mode coupling (SMC) model and the squeezed (Sqz) potential model discussed in the previous subsection.

4.3.3 Analysis of Tunneling Dynamics of the Excited $\tilde{A}$ State

Here the analysis of the experimental data shown in Table 4.6 done by Takada and Nakamura [31] is presented. This is based on knowledge of the calculated potential energy surface of the ground state $\tilde{X}$ mentioned in the previous subsection and the calculations with use of the three-dimensional model potential of Equation (4.23) [31].

First, let us consider the effects of ν'_{13} and ν'_{14}. Since the former has a_1 symmetry, the symmetry of the potential energy surface should be that of the symmetric mode coupling model [Equation (4.15)], in which x and y correspond to the direct tunneling coordinate and ν'_{13} mode, respectively. According to the numerical results shown in Table 4.1 and Table 4.8, the excitation of this ν'_{13} mode is thus expected to promote the tunneling. Experimentally, the first excitation of ν'_{13} gives the tunneling splitting of 33 cm^{-1} (3 cm^{-1}) compared to the experimentally observed splitting of 20 cm^{-1} (2 cm^{-1}) of the vibrationally ground state in the case of TRNOH (TRNOD) (see Table 4.6). Our prediction qualitatively agrees well with this experimental data. If we assume that ν'_{14} mode is similar to the ν''_{14}, it has both a_1 and b_2 components. The first excitation of ν'_{14} promotes the tunneling, which can be explained by the symmetric mode coupling (SMC) model. The excitation by two quanta in this mode of TRNOH, however, suppresses the tunneling slightly compared to the first excited state. This is probably because ν'_{14} has the b_2 component and the potential energy surface has the coupling to antisymmetric mode coupling (ASMC) type.

If we assume that the ν'_{11} mode has the same b_2 symmetry as the ν''_{11} mode, the symmetry of the potential energy should be that of antisymmetric mode coupling (ASMC) model [see Equation (4.18)]. Thus, the suppression of the tunneling by the ν'_{11} excitation can be understood by the out-of-phase cancellation in the Herring formula. Unfortunately, long excitation progression of this mode has not been observed. As emphasized in the previous subsection, excitation of this mode is expected to give an oscillatory behavior of the tunneling splitting as a function of the vibrational quantum number. This may be considered as an experimental observation of *mixed tunneling*.

In Table 4.6, the long progression of the ν'_{26} mode has been observed and the corresponding tunneling splitting decreases monotonically with its excitation. To know the symmetry of the potential energy surface, we need information on the normal mode. It has been repeatedly pointed out [75,76] that the strong Duschinsky effect between the ν_{25} mode and the ν_{26} mode is found in the $\tilde{X} - \tilde{A}$ electronic spectra. Namely, the ν'_{26} normal mode of $\tilde{A}$ state mainly consists of the mixture of the ν''_{25} and the ν''_{26} modes of $\tilde{X}$ state. Since the reliable MO calculations of the normal modes ν''_{25} and ν''_{26} are available (depicted schematically in Figure 4.8), the ν'_{26} mode can be guessed to include the antisymmetric wagging motion of two oxygen atoms. Therefore, the deviation of the ν'_{26} mode from planar structure makes the O–O distance larger and thus makes the barrier height of the proton tunneling between the two oxygen atoms larger. This clearly indicates the existence of squeezing effect in the potential energy surface. If we assume that the stable structure is planar in $\tilde{A}$ state, the potential energy surface must be symmetric with respect to the deviation of the ν'_{26} mode. Thus, the ν'_{26} mode coupled to the tunneling coordinate may be described by the squeezed model potential, in which x and y correspond to the direct tunneling mode and the ν'_{26} mode, respectively. It should be noted that ν'_{26} is the mode with the lowest frequency 39 cm^{-1}, while the direct tunneling mode has a frequency of about 3000 cm^{-1}. Thus, this case definitely fits the sudden regime. The numerical results for the squeezed model potential shown in Table 4.4 and the three-dimensional model for the $\tilde{X}$ state [Equation (4.23)] shown in Table 4.8 should qualitatively be in accordance with the experimental finding. In Table 4.6, the other two modes ν'_{19} and ν'_{25} with symmetries different from a_1 have the same tendency as that of ν'_{26}. This again indicates that the symmetry of coupling is important to clarify the effect of vibrational excitation on tunneling.

As was mentioned before, the present analysis is based on the assumption that the symmetry of the potential energy surface of $\tilde{A}$ state is similar to that of the ground state $\tilde{X}$ and that the stable structure of the excited state is planar. There have been carried out some quantum chemical calculations of the excited $\tilde{A}$ state [81,84–87]. However, the planarity of the equilibrium geometry of this state is not satisfactorily clarified yet [81,86,87]. Wojcik et al. [87] have obtained slightly nonplanar structure, but their analysis of tunneling splitting is in good accordance with the analysis mentioned above.

5 Nonadiabatic Tunneling

5.1 DEFINITION AND QUALITATIVE EXPLANATION

If two diabatic potential curves cross with opposite signs of slopes, then the lower and upper adiabatic potential curves have an attractive well and a potential barrier, respectively (see Figure 1.2). At energies lower than the top of the lower adiabatic potential, quantum mechanical tunneling can occur. This is called *nonadiabatic tunneling* and this type of potential curve crossing is called *nonadiabatic tunneling type* curve crossing (see Figure 1.2) [48]. This type of curve crossing plays important roles in chemical physics, since the so-called conical intersection of potential energy surfaces is ubiquitous in molecular systems [88]. A two-dimensional schematic view of conical intersection is drawn in Figure 5.1. The famous Jahn-Teller effect is related to this conical intersection [89–91]. When the potential energy surfaces are described by N independent coordinates, they can have a real potential surface crossing of $(N-2)$ dimensions, if the electronic symmetry of the two states is the same. This is called the Neumann-Wigner noncrossing rule. This can be simply proved as follows. Let $V_1(\mathbf{x})$, $V_2(\mathbf{x})$, and $V_{12}(\mathbf{x})$, be two diabatic potentials and the diabatic coupling between the two, where $\mathbf{x}$ collectively represents the relative nuclear coordinates. Then the corresponding adiabatic potentials $W_{1,2}(\mathbf{x})$ with $W_1(\mathbf{x}) \geq W_2(\mathbf{x})$ are given by

$$W_{1,2}(\mathbf{x}) = \frac{1}{2}\left[V_1(\mathbf{x}) + V_2(\mathbf{x}) \pm \sqrt{(V_1(\mathbf{x}) - V_2(\mathbf{x}))^2 + 4V_{12}(\mathbf{x})^2}\right]. \tag{5.1}$$

In order for the two adiabatic potentials $W_{1,2}(\mathbf{x})$ to cross, two terms in the square root should be zero at the same time that gives two conditions. Thus, the potential surface crossing is of $(N-2)$ dimensions or lower. In the case of a one-dimensional system, no crossing is possible. The two adiabatic curves can come close together but never cross on the real axis. This is called avoided crossing. In the case of $N = 2$, real crossing is possible at one point, namely, at the apex of two cones, at energy E_X as shown in Figure 5.1 (conical intersection). If the electronic symmetry of the two states is different, then there is no diabatic coupling, i.e., $V_{12}(\mathbf{x}) = 0$ and the crossing surface can be $(N-1)$ dimension. If we cut the two cones in Figure 5.1 by a vertical plane near the apex, then we can easily see that the two potential curves of the cross section are of the nonadiabatic tunneling type. The crossing of two diabatic potential curves with the same sign of slopes is called Landau-Zener type [48]. These two types of curve crossings present mathematically as well as physically quite different problems. The basic differential equations for the two cases show very different characteristics [48]. Physically, as is clear from the scheme of potential energy curves shown in Figure 1.2, the transmission—namely, the nonadiabatic tunneling—occurs in the case of nonadiabatic tunneling type. For both nonadiabatic tunneling type and the Landau-Zener type, the transition between the two adiabatic states is induced by

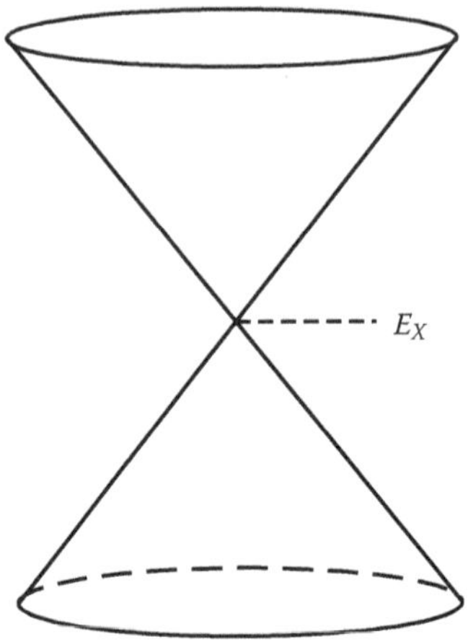

FIGURE 5.1 Schematic view of conical intersection of two adiabatic potential energy surfaces. Real crossing occurs at the energy E_X.

the nonadiabatic coupling, which is given by the following expression:

$$\text{Nonadiabatic coupling} \propto \frac{[\nabla_x V_{12}][V_1 - V_2] - V[\nabla_x(V_1 - V_2)]}{(W_1 - W_2)^2}\nabla_x, \tag{5.2}$$

which is a quantum mechanical operator. The transition is called nonadiabatic transition. Since the adiabatic states are defined as the eigenstates of the electronic Hamiltonian at fixed nuclear coordinates, this coupling comes from the nuclear motion, namely, the nuclear Laplacian operator, as is seen from the above expression.

It should be noticed that the nonadiabatic tunneling is affected by the nonadiabatic coupling even at energies lower than the top of the lower adiabatic potential. This is not widely recognized, but it is very dangerous to neglect this effect. It is easy to understand that no transmission is possible if there is no diabatic coupling, $V_{12}(\mathbf{x}) = 0$, between the two diabatic states. Figure 5.2 demonstrates this large difference between the ordinary tunneling and the nonadiabatic tunneling. This figure shows the transmission probabilities as a function of energy for the ordinary tunneling (monotonous curve) and the nonadiabatic tunneling (oscillatory curve). The oscillation appears at energies higher than the bottom of the upper adiabatic potential curve and the zero transmission occurs at certain discrete energies. This zero-transmission phenomenon is called *complete reflection* [48], which is a very unique phenomenon in nonadiabatic tunneling and is explained in more detail in the next section. The monotonically increasing curve represents the transmission probability in the case where the nonadiabatic coupling is neglected. The nonadiabatic tunneling probability is always smaller than the corresponding ordinary tunneling probability with the coupling neglected. As mentioned above, the nonadiabatic tunneling probability becomes zero when the diabatic coupling is zero. Namely, the difference between the two transmission probabilities depends on the coupling strength. Since this complete reflection phenomenon is very unique, various applications can be thought of [48]. If we have two same

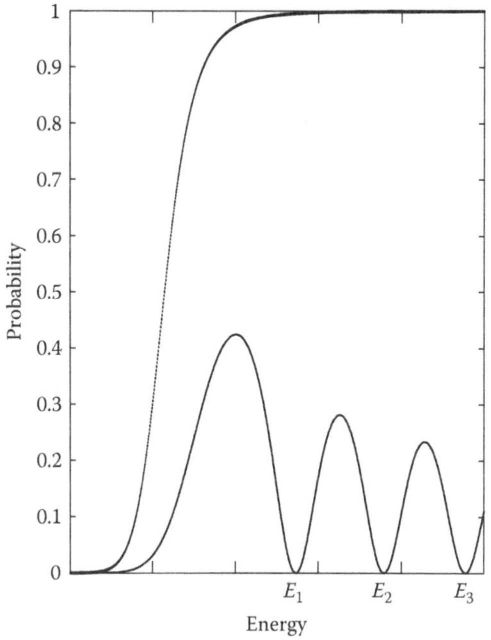

FIGURE 5.2 Transmission probability vs. energy. The oscillatory curve (solid line) corresponds to the nonadiabatic tunneling with the nonadiabatic coupling with the excited state taken into account. The monotonous curve (dotted line) is the ordinary transmission probability with the coupling neglected.

nonadiabatic tunneling type potential units, the wave can be trapped forever in between the two units at the energies of complete reflection. This means that we can create bound states in the continuum [92]. When we have periodic series of these potential units, complete transmission is possible at certain energies. If we can somehow click one of the potential units to make the complete transmission energy coincide with the complete reflection energy, then switching the transmission off and on—namely, molecular switching—becomes possible [92–94]. Theoretical application of this idea to realistic molecules has been considered by taking hydrogen transmission through a carbon ring as an example [95,96]. Finally, laser control of selective photodissociation also becomes possible. If an electronically excited state has a potential barrier, we can create two nonadiabatic tunneling type curve crossings by shining CW (continuous wave) laser and dressing up the attractive ground state (see Figure 5.3). By adjusting the laser frequency $\hbar\omega$, we can create complete reflection condition either at the right-side curve crossing for the dissociation to AB + C or at the left-side curve crossing for the dissociation to A + BC and break either one of the chemical bonds to dissociate the molecule selectively. Figure 5.3 depicts this control scheme schematically. This idea was applied to HI, HOD, and CH_3SH molecules [97,98].

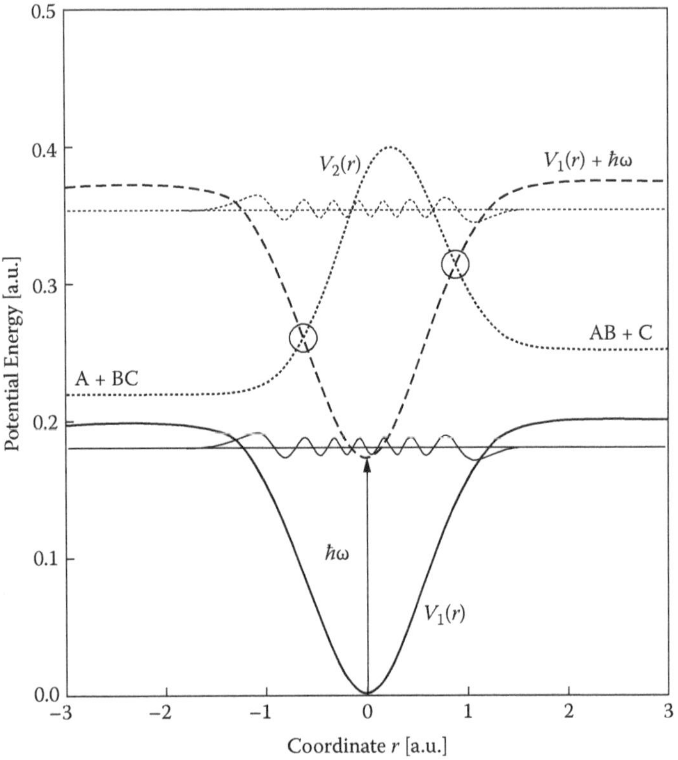

FIGURE 5.3 One-dimensional model of the control of photodissociation with use of the complete reflection phenomenon. Solid line: ground state $V_1(r)$; dotted line: excited state $V_2(r)$. Two circles represent the nonadiabatic tunneling type curve crossings between the excited state and the dressed ground state. A vibrationally excited state that dissociates either to the right or to the left is depicted. (Taken from Reference [97] with permission.)

5.2 ONE-DIMENSIONAL THEORY

For the problem of nonadiabatic tunneling transition, only one-dimensional theory is available. The famous Landau-Zener-Stückelberg theory was developed in 1932 [99–101]. The Landau-Zener formula for the nonadiabatic transition is well known. Unfortunately, however, the theory has some serious defects such that it cannot be used at energies near and lower than the crossing point, and many theoretical works were carried out to try to improve this [48]. Finally, Zhu and Nakamura succeeded in obtaining a whole set of semiclassical analytical formulas for transition amplitudes, including phases for both Landau-Zener type and the nonadiabatic tunneling type (Zhu-Nakamura theory) [102]. The formulas can be applied practically whatever the coupling strength and the energy are. The reviews are given in [103–107] and the whole set of formulas is summarized in the directly applicable way in the appendix of the review [108] and the book [48]. The formulas appropriate for the nonadiabatic tunneling type of potential crossing are given below.

The basic dimensionless parameters a^2 and b^2 to characterize the potential curve crossing are defined as (see Figure 5.4)

$$a^2 = \frac{(1-\gamma^2)\hbar^2}{m(R_b - R_t)^2(E_b - E_t)} \tag{5.3}$$

and

$$b^2 = \frac{E - (E_b + E_t)/2}{(E_b - E_t)/2}, \tag{5.4}$$

where m is the mass and

$$\gamma = \frac{E_b - E_t}{E_2([R_b + R_t]/2) - E_1([R_b + R_t]/2)}, \tag{5.5}$$

where $R_t(E_t)$ and $R_b(E_b)$ are the positions (energies) of the potential top and bottom, respectively. When $R_b = R_t$, $\gamma = 1$ and the following definition of a^2 should be used:

$$a^2 = \frac{\hbar^2}{4m(E_b - E_t)}\left[\frac{\partial^2 E_2(R)}{\partial R^2}|_{R=R_b} - \frac{\partial^2 E_1(R)}{\partial R^2}|_{R=R_t}\right]. \tag{5.6}$$

The above expressions are defined in terms of adiabatic potentials $E_1(R)$ and $E_2(R)$. The treatment in the adiabatic state representation is generally more accurate than the diabatic state representation. Since the diabatic state representation is, however, sometimes more convenient, the corresponding expressions in terms of the diabatic states are also given here for convenience:

$$a^2 = \frac{\hbar^2}{2m}\frac{F(F_1 - F_2)}{8V_X^3} \tag{5.7}$$

and

$$b^2 = (E - E_X)\frac{F_1 - F_2}{2FV_X}, \tag{5.8}$$

where $F_j (j = 1, 2)$ is the slope of the jth diabatic potential with $F = \sqrt{|F_1 F_2|}$, and V_X and E_X are the diabatic coupling and energy at the crossing point. Without loss of generality $F_1 - F_2$ can be assumed to be positive.

The total scattering matrix S in the two-state problem shown in Figure 5.4 is given by

$$S_{mn} = S^{\mathrm{R}}_{mn} \exp[i(\eta_m + \eta_n)], \tag{5.9}$$

where

$$\eta_1 = \lim_{R\to\infty}\left[\int_{T_1^r}^{R} K_1(R)dR - K_1(R)R + \frac{\pi}{4}\right], \tag{5.10}$$

$$\eta_2 = \lim_{R\to-\infty}\left[-\int_{T_1^l}^{R} K_1(R)dR + K_1(R)R + \frac{\pi}{4}\right], \tag{5.11}$$

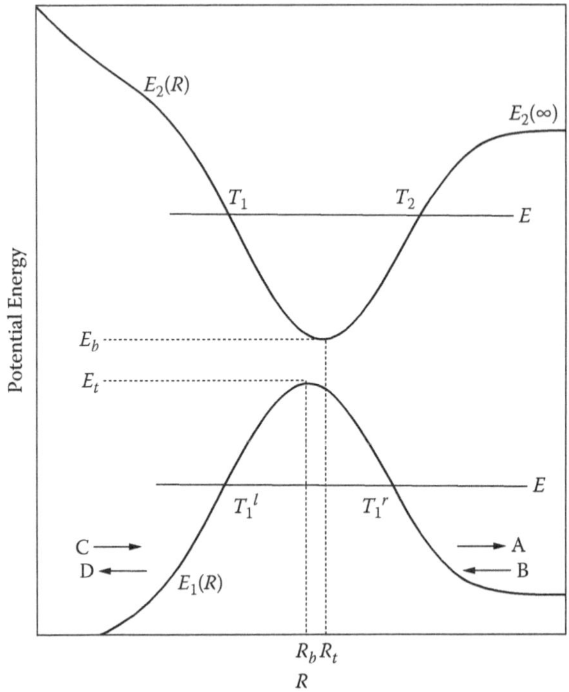

FIGURE 5.4 Nonadiabatic tunneling type potential curve crossing. $E_j(R)\,(j = 1, 2)$ is the jth adiabatic potential curve. Various notations are used in the text. (Taken from Reference [48] with permission.)

$$S^{\mathrm{R}} = \frac{1}{1 + U_1 U_2} \begin{pmatrix} e^{i\Delta_{11}} & U_2 e^{i\Delta_{12}} \\ U_2 e^{i\Delta_{12}} & e^{i\Delta_{22}} \end{pmatrix}, \tag{5.12}$$

$$K_j(R) = \sqrt{\frac{2m}{\hbar^2}[E - E_j(R)]} \ \text{for} \ (j = 1, 2), \tag{5.13}$$

and

$$U_2 = 2i\ \mathrm{Im}(U_1)/(|U_1|^2 - 1), \tag{5.14}$$

where U_1 and U_2 are the so-called Stokes constants [48]. Note that T_1^l and T_1^r should be replaced by R_t at $E \geq E_b$, and the indices 1 and 2 of the S-matrix correspond to the right and left side of the barrier, respectively. Namely, the diagonal (off-diagonal) elements of the S-matrix represent reflection (transmission). The S^R-matrix is called reduced scattering matrix and represents the transition amplitudes in the transition zone $T_1^l \leq R \leq T_1^r$. The other quantities in Equation (5.12) are defined in the next subsection. The transfer matrix N at $E < E_2(\infty)$ is given by

$$\begin{pmatrix} C \\ D \end{pmatrix} = N \begin{pmatrix} A \\ B \end{pmatrix}, \tag{5.15}$$

where $A(B)$ is the coefficient of the wave running to the right (left) along $E_1(R)$ on the right side of the barrier, and $C(D)$ is the coefficient of the wave running to the right (left) along $E_1(R)$ on the left side of the barrier. The N-matrix is related to the S^{R}-matrix as

$$N_{11} = 1/S^{\mathrm{R}}_{12}, \qquad N_{12} = S^{\mathrm{R}}_{22}/S^{\mathrm{R}}_{12}, \qquad N_{21} = N^*_{12}, \qquad N_{22} = N^*_{11}. \tag{5.16}$$

The Zhu-Nakamura theory provides analytical expressions separately for the following three energy regions: $E \leq E_t$ (top of the lower adiabatic potential), $E_t \leq E \leq E_b$ (bottom of the upper adiabatic potential), and $E_b \leq E$. The formulas given here contain some empirical corrections so that the formulas can cover even those small regions of parameters in which the original formulas are not necessarily very accurate. Thus the formulas given below can be directly applied to practical problems.

5.2.1 Case of $E \leq E_t$

The required quantities in Equation (5.12) are given by

$$\Delta_{12} = \Delta_{11} = \Delta_{22} = -2\sigma_{\mathrm{ZN}}, \tag{5.17}$$

$$\mathrm{Re}\, U_1 = \sin(2\sigma_c)\left\{\frac{0.5\sqrt{a^2}}{1+\sqrt{a^2}}\sqrt{B\left(\frac{\sigma_c}{\pi}\right)}e^{-\delta_{\mathrm{ZN}}} + \frac{e^{\delta_{\mathrm{ZN}}}}{\sqrt{B(\sigma_c/\pi)}}\right\}, \tag{5.18}$$

$$\mathrm{Im}\, U_1 = \cos(2\sigma_c)\sqrt{\frac{(\mathrm{Re}\, U_1)^2}{\sin^2(2\sigma_c)} + \frac{1}{\cos^2(2\sigma_c)} - \frac{1}{2\sin(\sigma_c)}\left|\frac{\mathrm{Re}\, U_1}{\cos(\sigma_c)}\right|}, \tag{5.19}$$

$$\sigma_c = \sigma_{\mathrm{ZN}}(1 - 0.32 \times 10^{-2/a^2} e^{-\delta_{\mathrm{ZN}}}), \tag{5.20}$$

$$\delta_{\mathrm{ZN}} = \int_{T_1^l}^{T_1^r} |K_1(R)| dR, \tag{5.21}$$

and

$$\sigma_{\mathrm{ZN}} = \frac{\pi}{8a|b|}\frac{1}{2}\frac{\sqrt{6+10\sqrt{1-1/b^4}}}{1+\sqrt{1-1/b^4}}, \tag{5.22}$$

where the function $B(x)$ is given by

$$B(x) = \frac{2\pi x^{2x} e^{-2x}}{x\Gamma^2(x)}. \tag{5.23}$$

The physically meaningful overall transmission probability, namely, the nonadiabatic tunneling probability, is given by

$$P_{12} = \frac{B(\sigma_c/\pi)e^{-2\delta_{\mathrm{ZN}}}}{[1 + (0.5\sqrt{a^2}/[\sqrt{a^2}+1])B(\sigma_c/\pi)e^{-2\delta_{\mathrm{ZN}}}]^2 + B(\sigma_c/\pi)e^{-2\delta_{\mathrm{ZN}}}}. \tag{5.24}$$

It should be noted that when $a^2 \to 0$ (large diabatic coupling limit), we have

$$P_{12} = \frac{e^{-2\delta_{\mathrm{ZN}}}}{1 + e^{-2\delta_{\mathrm{ZN}}}}, \tag{5.25}$$

which agrees with the ordinary single potential barrier penetration probability. It should be noted that the effect of nonadiabatic coupling is represented by σ_c in Equation (5.24) and that this effect cannot be separated out. Nonadiabatic coupling effect and tunneling are coupled.

5.2.2 Case of $E_t \leq E \leq E_b$

The necessary quantities to define S^R in Equation (5.12) are given by

$$\Delta_{12} = \sigma_{ZN}, \qquad \Delta_{11} = \sigma_{ZN} - 2\sigma_0^{ZN}, \quad \text{and} \quad \Delta_{22} = \sigma_{ZN} + 2\sigma_0^{ZN}, \tag{5.26}$$

$$\sigma_0^{ZN} = -\frac{1}{3}(R_t - R_b)K_1(R_t)(1+b^2), \tag{5.27}$$

$$U_1 = i[\sqrt{1+W^2}e^{i\phi} - 1]/W, \tag{5.28}$$

$$\phi = \sigma_{ZN} + \arg\Gamma\left(\frac{1}{2} + i\frac{\delta_{ZN}}{\pi}\right) - \frac{\delta_{ZN}}{\pi}\ln\left(\frac{\delta_{ZN}}{\pi}\right) + \frac{\delta_{ZN}}{\pi} - h_2, \tag{5.29}$$

$$h_2 = 0.34\frac{a^{0.7}(a^{0.7}+0.35)}{a^{2.1}+0.73}(0.42+b^2)\left(2+\frac{100b^2}{100+a^2}\right)^{0.25}, \tag{5.30}$$

$$\sigma_{ZN} = -\frac{1}{\sqrt{a^2}}\left[0.057(1+b^2)^{0.25} + \frac{1}{3}\right](1-b^2)\sqrt{5+3b^2}, \tag{5.31}$$

$$\delta_{ZN} = \frac{1}{\sqrt{a^2}}\left[0.057(1-b^2)^{0.25} + \frac{1}{3}\right](1+b^2)\sqrt{5-3b^2}, \tag{5.32}$$

$$W = \frac{h_3}{a^{2/3}}\int_0^\infty \cos\left[\frac{t^3}{3} - \frac{b^2}{a^{2/3}}t - \frac{h_4}{a^{2/3}}\frac{t}{h_5 + a^{1/3}t}\right]dt, \tag{5.33}$$

$$h_3 = 1 + \frac{0.38}{a^2}(1+b^2)^{1.2-0.4b^2}, \tag{5.34}$$

$$h_4 = \frac{\sqrt{a^2}-3b^2}{\sqrt{a^2}+3}\sqrt{1.23+b^2}, \tag{5.35}$$

$$h_5 = 0.61\sqrt{2+b^2}. \tag{5.36}$$

The physically meaningful overall transmission probability is given by

$$P_{12} = \frac{W^2}{1+W^2}. \tag{5.37}$$

5.2.3 Case of $E_b \leq E$

The various quantities in Equation (5.12) are given as follows:

$$\Delta_{12} = \sigma_{ZN}, \tag{5.38}$$

$$\Delta_{11} = 2\int_{T_1}^{R_b} K_2(R)dR - 2\sigma_0^{ZN}, \tag{5.39}$$

$$\Delta_{22} = 2\int_{R_b}^{T_2} K_2(R)dR + 2\sigma_0^{ZN}, \tag{5.40}$$

$$\sigma_0^{ZN} = \left(\frac{R_b - R_t}{2}\right)\left\{K_1(R_t) + K_2(R_b) + \frac{1}{3}\frac{[K_1(R_t) - K_2(R_b)]^2}{K_1(R_t) + K_2(R_b)}\right\}, \tag{5.41}$$

$$U_1 = i\sqrt{1 - p_{\mathrm{ZN}}}\exp[i(\sigma_{\mathrm{ZN}} - \bar{\phi}_{\mathrm{S}})], \tag{5.42}$$

$$\sigma_{\mathrm{ZN}} = \int_{T_1}^{T_2} K_2(R)dR, \tag{5.43}$$

and

$$\delta_{\mathrm{ZN}} = \frac{\pi}{8ab}\frac{1}{2}\frac{\sqrt{6 + 10\sqrt{1 - 1/b^4}}}{1 + \sqrt{1 - 1/b^4}}. \tag{5.44}$$

The nonadiabatic transition probability p_{ZN} for one passage of the crossing point is given by

$$p_{\mathrm{ZN}} = \exp\left[-\frac{\pi}{4ab}\left(\frac{2}{1 + \sqrt{1 - b^{-4}(0.72 - 0.62a^{1.43})}}\right)^{1/2}\right], \tag{5.45}$$

and the overall transmission probability is given by

$$P_{12} = \frac{4\cos^2(\sigma_{\mathrm{ZN}} - \bar{\phi}_{\mathrm{S}})}{4\cos^2(\sigma_{\mathrm{ZN}} - \bar{\phi}_{\mathrm{S}}) + (p_{\mathrm{ZN}})^2/(1 - p_{\mathrm{ZN}})}, \tag{5.46}$$

where

$$\bar{\phi}_S = \phi_S + h_1, \tag{5.47}$$

$$h_1 = \frac{0.23a^{1/2}}{a^{1/2} + 0.75}40^{-\sigma_{ZN}}, \tag{5.48}$$

$$\phi_{\mathrm{S}} = -\frac{\delta_\psi}{\pi} + \frac{\delta_\psi}{\pi}\ln\left(\frac{\delta_\psi}{\pi}\right) - \arg\Gamma\left(i\frac{\delta_\psi}{\pi}\right) - \frac{\pi}{4}, \tag{5.49}$$

and

$$\delta_\psi = \delta_{ZN}\left(1 + \frac{5a^{1/2}}{a^{1/2} + 0.8}10^{-\sigma_{ZN}}\right). \tag{5.50}$$

The nonadiabatic transition matrix I, which represents the nonadiabatic transition at the avoided crossing point, is given by

$$I = \begin{pmatrix} \sqrt{1 - p_{\mathrm{ZN}}}e^{i\phi_{\mathrm{S}}} & \sqrt{p_{\mathrm{ZN}}}e^{i\sigma_0^{\mathrm{ZN}}} \\ -\sqrt{p_{\mathrm{ZN}}}e^{-i\sigma_0^{\mathrm{ZN}}} & \sqrt{1 - p_{\mathrm{ZN}}}e^{-i\phi_{\mathrm{S}}} \end{pmatrix}, \tag{5.51}$$

where p_{ZN} is given by Equation (5.45) and ϕ_S is the same as Equation (5.49). At $E > E_2(\infty)$, there is no turning point T_2, and σ_{ZN} in Equation (5.50) cannot be defined. Since this is just a small correction, we can use δ_{ZN} instead of δ_ψ.

It should be noted that when $\psi_{ZN} = \sigma_{ZN} - \bar{\phi}_S = (n + 1/2)\pi$ $(n = 0, 1, 2, \ldots)$, P_{12} of Equation (5.46) becomes zero. Namely, the intriguing phenomenon of *complete reflection* occurs, as was mentioned in the previous section. This occurs whatever the probability p_{ZN} is, i.e., whatever the potential shapes are. The complete reflection condition is somewhat like the Bohr-Sommerfeld quantization condition but contains the effect of nonadiabatic coupling as represented by the quantity $\bar{\phi}_S$. This phenomenon is due to the destructive phase interference at the exit between the wave that transmits through the potential system without trapping and the wave that comes out after being trapped by the upper adiabatic potential.

The above Zhu-Nakamura formulas have been numerically tested to work well for various cases in comparison with quantum mechanically exact numerical calculations [48,102–107]. Figure 5.5 shows one example (see (6) of Reference [102]). The model potentials used are given by (in atomic units)

$$V_1(x) = 0.5e^{7.5x},$$

$$V_2(x) = 0.5e^{-3.75x},$$

$$V_{12}(x) = \frac{3}{2\sqrt{2}} V_0 e^{-2.5x^2}. \qquad (5.52)$$

The reduced mass is taken to be 1000 a.u. and the coupling parameter V_0 is varied so that the parameter a^2 calculated by Equation (5.7) covers the three regimes: $a^2 = 10.0$ (weak diabatic coupling), $a^2 = 1.0$ (intermediate coupling), and $a^2 = 0.1$ (strong coupling). Figures 5.5(a)–(c) correspond to strong, intermediate, and weak coupling cases, respectively. Solid and dashed lines represent the results of the Zhu-Nakamura formulas with use of the parameters a^2 and b^2 in the adiabatic state and diabatic state representation, respectively. These agree well with the exact numerical results depicted by dots. The fact that the adiabatic state representation works better than the diabatic state representation can be seen clearly in the localizability of the transition and the phases of S-matrix [48,102–107].

The Zhu-Nakamura formulas can also be applied to multichannel problems, since the nonadiabatic transition is well localized at avoided crossing in the adiabatic state representation [48,109]. This was also tested for various potential systems. One numerical example is depicted in Figure 5.6 with use of the potential system shown in Figure 5.7 (see (3) of Reference [109]). These potentials are given as follows (in atomic units):

$$V_1(R) = 0.037e^{-1.3(R-3.25)} - 0.034,$$

$$V_2(R) = 0.037e^{-1.3(R-3.25)} - 0.012,$$

$$V_3(R) = 0.4057\left[1 - e^{-0.344(R-3)}\right]^2 - 0.03,$$

$$V_4(R) = 0.4057\left[1 - e^{-0.344(R-3)}\right]^2. \qquad (5.53)$$

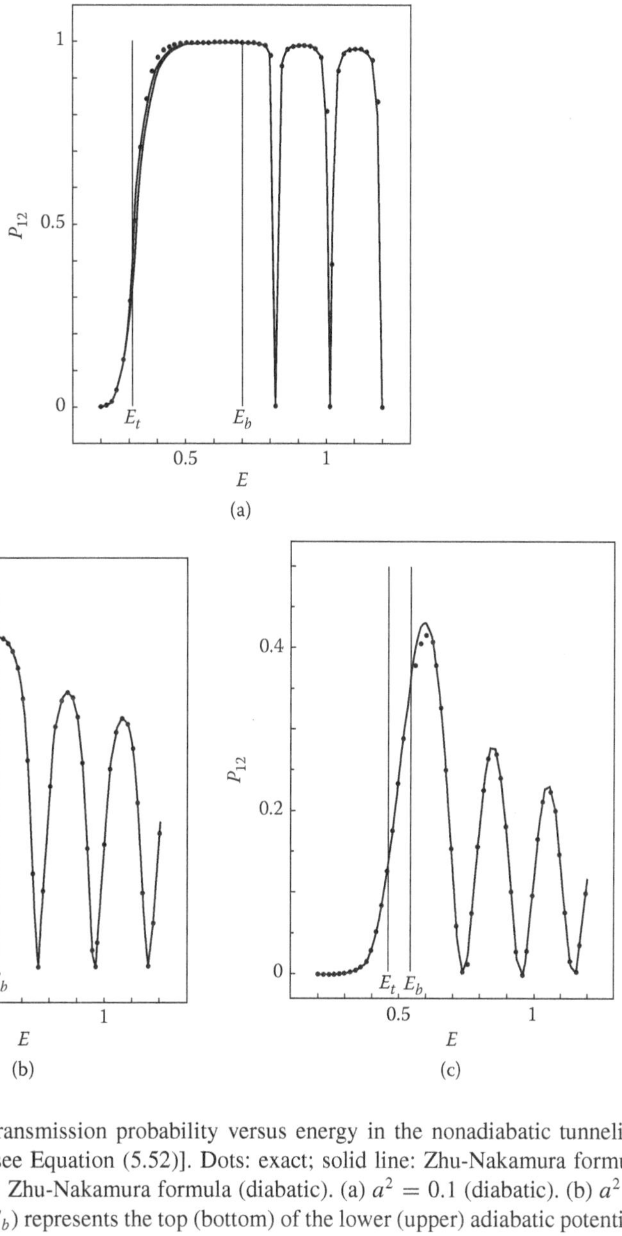

FIGURE 5.5 Overall transmission probability versus energy in the nonadiabatic tunneling type potential system [see Equation (5.52)]. Dots: exact; solid line: Zhu-Nakamura formula (adiabatic); dashed line: Zhu-Nakamura formula (diabatic). (a) $a^2 = 0.1$ (diabatic). (b) $a^2 = 1.0$. (c) $a^2 = 10.0$. $E_t(E_b)$ represents the top (bottom) of the lower (upper) adiabatic potential. (Taken from Reference (6) of [102] with permission.)

Diabatic couplings are given as

$$V_{13}(R) = V_{14}(R) = V_{23}(R) = V_{24}(R) = \frac{0.004}{1 + e^{R-3}},$$
$$V_{12}(R) = V_{34}(R) = 0. \tag{5.54}$$

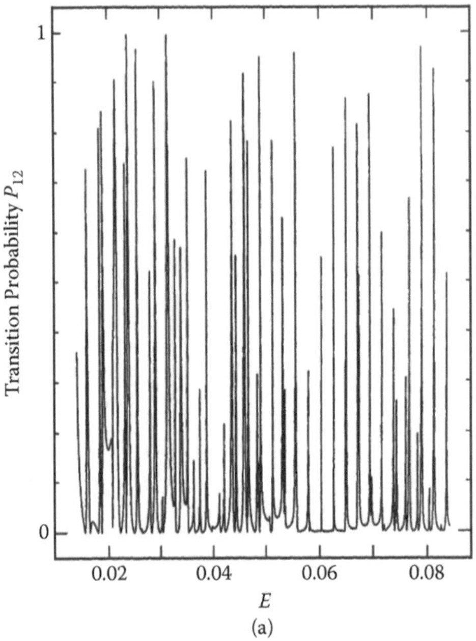

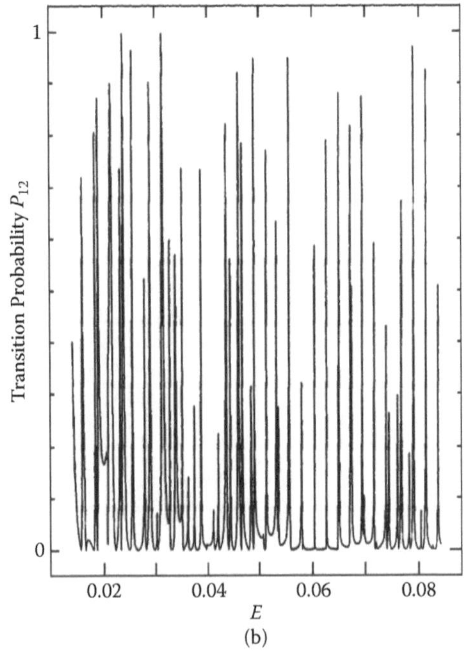

FIGURE 5.6 Overall transition probability P_{12} versus energy for the potential system given in Figure 5.7. (a) Quantum mechanical exact numerical solution of coupled equations. (b) Semiclassical results with use of the Zhu-Nakamura formulas. The starting energy and the energy step used in the computation are exactly the same for the two cases. (Taken from Reference (8) of [102] with permission.)

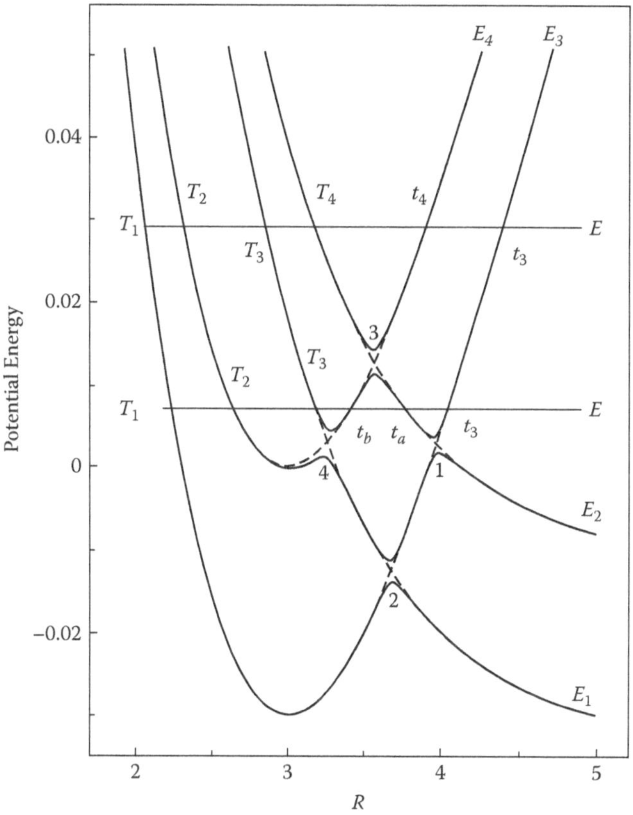

FIGURE 5.7 Four-state potential curve diagram of Equations (5.53) and (5.54). (Taken from Reference (8) of [102] with permission.)

The reduced mass used is that of an oxygen molecule ($m = 29377.3$). This model system was taken from some states of O_2 and the coupling represents the spin-orbit interaction among vibrational states of the oxygen molecule [110]. The coupling strength chosen is such a value that any perturbation theory does not work at all. The parameters a^2 [see Equation (5.3)] for the four avoided crossings are 1.8, 3.61, 4.5, and 7.88, which are out of the range in which the perturbation theory works ($a^2 \sim 10$). Figure 5.6(a) is the result of quantum mechanical numerical solution of coupled differential equations and Figure 5.6(b) is the corresponding semiclassical result with use of the Zhu-Nakamura formulas. The starting energy and the energy step of the calculations are taken to be exactly the same for both cases. There is excellent agreement between the two results for a wide range of energy. Even the very detailed resonance structures are well reproduced by the semiclassical Zhu-Nakamura theory.

Multidimensional theory is not yet available, unfortunately, not only for the nonadiabatic tunneling problem but also for general nonadiabatic transition problems. For practical applications, the Zhu-Nakamura formulas for transition amplitude including phases can be incorporated into classical or semiclassical propagation schemes. These are ZN-TSH method [108,111,112] in which the Zhu-Nakamura formulas of transition probability are incorporated into the trajectory surface hopping (TSH) method [113–117] and ZN-HKSCIVR method [108,112,118] in which the Zhu-Nakamura transition amplitude is incorporated into the Herman-Kluk type semiclassical initial value representation method [119–121]. We don't go into details here. Those who are interested may refer to the above references.

6 Multidimensional Theory of Tunneling Splitting

6.1 GENERAL FORMULATION

In this section, generalizations of the instanton theory and the modified WKB theory described in Chapter 2 to multidimensional space are presented [43]. Those who are not interested in the generalization of the instanton approach and the proof of its equivalence to the modified WKB theory can skip Section 6.1.1 and Section 6.1.2. In Section 6.1.3 a general WKB formulation for a general Hamiltonian in curved space is provided and its final expression of tunneling splitting can be directly applied to any real systems.

In this section the mass scaled coordinates are used and the Planck constant $\hbar$ is assumed to be unity, unless otherwise explicitly mentioned. The mass m and the Planck constant $\hbar$ can be recovered in all the equations and final results by making the following substitutions. This applies also to other sections in this chapter.

$$\mathbf{q} \to \sqrt{m}\mathbf{q}, \quad \frac{\partial}{\partial \mathbf{q}} \to \frac{1}{\sqrt{m}} \frac{\partial}{\partial \mathbf{q}} \quad (\mathbf{q} : \text{coordinate}), \tag{6.1}$$

$$\mathbf{p} \to \frac{1}{\sqrt{m}}\mathbf{p} \quad (\mathbf{p} : \text{momentum}), \tag{6.2}$$

$$V_{\mathbf{qq}} \to \frac{1}{m} V_{\mathbf{qq}} \quad (V : \text{potential}), \tag{6.3}$$

$$\mathbf{A} \to \frac{1}{m}\mathbf{A}, \tag{6.4}$$

where $A_{ij} = \partial W_0 / \partial q_i \partial q_j$ [see Equations (6.74), (6.96), (6.155), and (6.181)], and

$$S \to \frac{1}{\hbar} S \quad (\mathrm{S} : \text{action}) \tag{6.5}$$

and

$$\Delta \to \sqrt{\hbar}\Delta \quad (\Delta : \text{tunneling splitting}). \tag{6.6}$$

6.1.1 Multidimensional Extension of the Instanton Theory

To evaluate the N-dimensional analog of the pre-exponential factor B [see Equation (2.135)], we consider the path integral over harmonic fluctuations along the instanton,

$$\int D\mathbf{q} \exp\left[-\int_{-\infty}^{\infty} \left(\frac{1}{2}\dot{\mathbf{q}}^2 + \sum_{ij} \frac{1}{2} V''_{ij}(\mathbf{q}_0(\tau) q_i q_j \right) d\tau \right], \tag{6.7}$$

where V''_{ij} is the matrix of second derivatives $\partial^2 V/\partial q^i \partial q^j$ and the integration is carried out over trajectories subjected to the condition $\mathbf{q}(\pm\infty) = 0$. Similarly to the one-dimensional case, the operator $\sum_{ij}[-\frac{\partial^2}{\partial\tau^2}\delta_{ij} + V''_{ij}(\mathbf{q}_0(\tau))]$ has a zero eigenvalue that manifests the invariance of the classical action with respect to the time shift of the instanton. We assume that this is the only zero eigenvalue that corresponds to the zero mode. To proceed with the multidimensional case, we first formally consider the above equation as it is and show that the functional integral can be factorized into two parts. The part that contains the singularity is purely one dimensional and its contribution to Δ_0 is actually given by Equation (2.135). The second part is convergent and leads to an extra pre-exponential factor that we calculate separately.

Let S parametrize the instanton trajectory $\mathbf{q}_0(S)$ and let $\{\xi_\alpha\}$ be a set of $N - 1$ transverse coordinates defined by the relation

$$Q_i = q_{0i}(S) + \sum_\alpha \tau_{i\alpha}(S)\xi_\alpha, \tag{6.8}$$

where $\mathbf{Q}$ is used to distinguish from $\mathbf{q}(\tau)$ in the functional integral. We choose the matrices $\tau_{i\alpha}(S)$ $(\alpha = 1, 2, \ldots, N - 1)$ and $\tau_{iN}(S) \equiv d\mathbf{q}_{0i}/dS$ to satisfy the relations

$$\sum_{i=1}^{N} \tau_{i\alpha}(S)\tau_{i\beta}(S) = \delta_{\alpha\beta}, \quad (\alpha, \beta = 1, 2, \ldots, N - 1) \tag{6.9}$$

and

$$\sum_{i=1}^{N} \tau_{i\alpha}(S)\frac{d\tau_{i\beta}}{dS} = 0, \quad (\alpha, \beta = 1, 2, \ldots, N - 1). \tag{6.10}$$

Hereafter the summation is taken from 1 to N for Latin indices and from 1 to $N - 1$ for the Greek ones. According to Equation (6.9), $\{\tau_{i\alpha}\}$ determine the orthogonal transformation to the local frame of reference. One can check that the second condition, Equation (6.10), makes the metric tensor diagonal in the coordinates S and ξ_α. This condition is not essential but it simplifies the following derivations. The instanton trajectory approaches the potential minima along one of the normal modes, which we numerate as the Nth. We can also choose $\{\xi_\alpha\}$ to coincide with the other $N - 1$ normal coordinates at the potential minimum. Due to Equation (6.10), we have

$$\frac{\tau_{i\alpha}(S)}{dS} = C_\alpha(S)\tau_{iN}(S), \tag{6.11}$$

where the quantities $C_\alpha(\alpha = 1, 2, \ldots, N - 1)$ define the curvature of $\mathbf{q}_0(S)$.

Let us now consider the classical mechanics in the new coordinates. We consider the classical motion in the upside-down potential and thus the Lagrangian reads

$$L = \frac{1}{2}\left(1 + \sum_\alpha C_\alpha \xi_\alpha\right)^2 \dot{S}^2 + \frac{1}{2}\sum_\alpha \dot{\xi}_\alpha^2 + V(S, \{\xi\}). \tag{6.12}$$

Since we are interested in the small deviations from the instanton trajectory, the potential is expanded as

$$V(S, \{\xi\}) = V_0(S) + \sum_\alpha a_\alpha(S)\xi_\alpha + \frac{1}{2}\sum_{\alpha\beta} b_{\alpha\beta}(S)\xi_\alpha\xi_\beta + o(\xi^2). \tag{6.13}$$

This gives the classical equations of motion for the instanton $[S_0(\tau), \xi_0(\tau) = 0]$ as

$$\ddot{S}_0(\tau) = \frac{dV_0}{dS} \text{ and } C_{0\alpha}\dot{S}_0^2 + a_{0\alpha} = 0, \tag{6.14}$$

where the index 0 indicates that all the quantities are taken at the instanton. For instance, $C_{0\alpha} \equiv C_\alpha[S_0(\tau)]$. Substituting $S = S_0 + s$ and expanding the Lagrangian L up to the second order terms s^2, ξ^2, and $s\xi$, we obtain

$$L = L^0 + \delta^1 L + \delta^2 L + \ldots, \tag{6.15}$$

where $L^0 = L(S_0, \dot{S}_0, \xi = 0)$, $\delta^1 L$ is zero due to the classical equation of motion, Equation (6.14), and we have

$$\delta^2 L(s, \dot{s}, \xi, \dot{\xi}, \tau) = \frac{1}{2}\dot{s}^2 + \frac{1}{2}\sum_\alpha \dot{\xi}_\alpha^2 + \dot{S}_0^2 s \sum_\alpha C'_{0\alpha}\xi_\alpha + 2\dot{S}_0\dot{s}\sum_\alpha C_{0\alpha}\xi_\alpha$$
$$+\frac{1}{2}\dot{S}_0^2\sum_\alpha C_{0\alpha}\xi_\alpha^2 + \frac{1}{2}V_0'' s^2 + s\sum_\alpha a'_{0\alpha}\xi_\alpha + \frac{1}{2}\sum_{\alpha,\beta} b_{0\alpha\beta}\xi_\alpha\xi_\beta. \tag{6.16}$$

Hereafter the prime stands for the derivative d/dS. It is easy to see that q_i and ξ_α (with $\xi_N = s$) are related by the orthogonal transformation

$$q_i = \sum_\alpha \tau_{i\alpha}(S_0(\tau))\xi_\alpha, \tag{6.17}$$

and the Lagrangian $\sum_i \dot{q}_i^2/2 + \sum_{ij}(\partial^2 V[\mathbf{q}_0(\tau)]/\partial q_i \partial q_j)q_i q_j/2$ coincides with $\delta^2 L(s, \{\xi\})$. Since the Jacobian $D\mathbf{q}/D\xi$ is unity, we can rewrite the functional integral, Equation (6.7), as

$$\int DsD\{\xi\}\exp\left[-\int_{-\infty}^{\infty}\delta^2 L(s, \dot{s}, \{\xi\}, \{\dot{\xi}\}, \tau)d\tau\right], \tag{6.18}$$

which is taken again over trajectories with zero boundary conditions. Now we perform the second transformation, which is just a shift in $s(\tau)$:

$$s(\tau) = u(\tau) + \dot{S}_0(\tau)v(\tau), \tag{6.19}$$

where $v(\tau)$ is related to $\{\xi(\tau)\}$ by

$$\dot{v}(\tau) = -2\sum_\alpha C_{0\alpha}(\tau)\xi_\alpha(\tau). \tag{6.20}$$

We note that the Jacobian $D(s, \xi)/D(u, \xi)$ is unity again and $u(\tau)$ satisfies the same zero boundary conditions as $s(\tau)$. The integral Equation (6.18) becomes

$$\int DuD\{\xi\}\exp\left[-\int_{-\infty}^{\infty}\delta^2 L(u, \dot{u}, \{\xi\}, \{\dot{\xi}\}, \tau)d\tau\right], \tag{6.21}$$

where the Lagrangian in the new coordinates $(u, \{\xi\})$ is given by

$$\delta^2 L(u, \dot{u}, \{\xi\}, \{\dot{\xi}\}, \tau) = \frac{1}{2}\dot{u}^2 + \frac{1}{2}V_0''u^2 + \frac{1}{2}\sum_{\alpha}\dot{\xi}_\alpha^2$$
$$+\frac{1}{2}\sum_{\alpha\beta}[b_{0\alpha\beta} - 3\dot{S}_0^2 C_{0\alpha}C_{0\beta}]\xi_\alpha\xi_\beta$$
$$+\frac{1}{2}\ddot{S}_0^2 v^2 - 2\dot{S}_0\ddot{S}_0 v\sum_{\alpha} C_{0\alpha}\xi_\alpha + \frac{1}{2}V_0''\dot{S}_0^2 v^2$$
$$+\ddot{S}_0\dot{u}v - 2\ddot{S}_0 u\sum_{\alpha} C_{0\alpha}\xi_\alpha + V_0''\dot{S}_0 uv. \qquad (6.22)$$

Using Equation (6.20) and the classical equations of motion, we can rewrite the third and fourth lines as

$$\text{third line} = \frac{1}{2}\ddot{S}_0^2 v^2 + \ddot{S}_0\dot{S}_0 v\dot{v} + \frac{1}{2}\left(\frac{d}{d\tau}\ddot{S}_0\right)\dot{S}_0 v^2 = \frac{1}{2}\frac{d}{d\tau}(\ddot{S}_0\dot{S}_0 v^2) \qquad (6.23)$$

and

$$\text{fourth line} = \ddot{S}_0\dot{u}v + \ddot{S}_0 u\dot{v} + \left(\frac{d}{d\tau}\ddot{S}_0\right)uv = \frac{d}{d\tau}(\ddot{S}_0 uv). \qquad (6.24)$$

Thus both terms do not contribute to the action and the functional integral breaks up into two factors:

$$\int Du \exp\left[-\int_{-\infty}^{\infty}\delta^2 L_u(u, \dot{u}, \tau)d\tau\right]\cdot\int D\{\xi\}\exp\left[-\int_{-\infty}^{\infty}\delta^2 L_\xi(\{\xi\}, \{\dot{\xi}\}, \tau)d\tau\right]. \qquad (6.25)$$

Here $\delta^2 L_u$ and $\delta^2 L_\xi$ are given by the first and second lines of Equation (6.22), respectively. Now we arrive at the following expression for the pre-exponential factor in the multidimensional case:

$$B = \sqrt{\frac{2S_0}{\pi}}\left[\frac{\det'\left(-\frac{\partial^2}{\partial\tau^2} + V_0''(S_0(\tau))\right)}{\det(-\frac{\partial^2}{\partial\tau^2} + \omega_N^2)}\right]^{-1/2}\left[\frac{\det\left(-\frac{\partial^2}{\partial\tau^2}\mathbf{1} + \Omega^2(\tau)\right)}{\det(-\frac{\partial^2}{\partial\tau^2}\mathbf{1} + \Omega_0^2)}\right]^{-1/2}. \qquad (6.26)$$

Here we have introduced the $(N-1)$-dimensional symmetric matrix $\Omega^2(\tau)$,

$$\Omega_{\alpha\beta}^2(\tau) = b_{0\alpha\beta}(\tau) - 3\dot{S}_0^2 C_\alpha(S_0(\tau))\, C_\beta(S_0(\tau)) \qquad (6.27)$$

and

$$(\Omega_0)_{\alpha\beta} = \delta_{\alpha\beta}\omega_\alpha, \qquad (6.28)$$

where ω_α are $N-1$ frequencies of the normal modes orthogonal to the instanton at potential minimum $\mathbf{q}_m$. The contribution of the one-dimensional part, i.e., the first two terms of Equation (6.26), was analyzed before in Chapter 2 and now we proceed with evaluation of the transverse factor, the third term of Equation (6.26).

Since the zero mode that indicates the translational invariance with respect to the time shift appears only in the one-dimensional part, all the eigenvalues of the transverse operator (with Dirichlet boundary conditions) $[-\partial^2/\partial\tau^2\mathbf{1} + \boldsymbol{\Omega}^2(\tau)]$ are positive. Let us first consider the simplest two-dimensional case in which $\Omega(\tau)$ is symmetric, $\Omega(-\tau) = \Omega(\tau)$. The instanton trajectory is centered at $\tau = 0$. Asymptotically $\Omega(\tau) \to \Omega_0$ and it differs significantly only in a finite time interval around $\tau = 0$. In order to evaluate the ratio of the determinants,

$$\frac{\det\left(-\frac{\partial^2}{\partial\tau^2} + \Omega^2(\tau)\right)}{\det\left(-\frac{\partial^2}{\partial\tau^2} + \Omega_0^2\right)}, \tag{6.29}$$

we proceed in the same way as in the one-dimensional case [see Equations (2.110) and (2.115)]. Then we need to evaluate

$$\frac{J(T/2)}{J_0(T/2)} = 2\Omega_0 J(T/2)\exp[-\Omega_0 T], \tag{6.30}$$

where the Jacobi field $J(T/2)$ satisfies the Jacobi equation,

$$\ddot{J}(\tau) = \Omega^2(\tau)J(\tau) \tag{6.31}$$

with the initial conditions [see Equations (2.113) and (2.114)],

$$J(-T/2) = 0, \quad \dot{J}(-T/2) = 1. \tag{6.32}$$

Let us introduce another solution $\eta(\tau)$ of the Jacobi equation such that

$$\eta(-T/2) = 1, \quad \dot{\eta}(-T/2) = \Omega_0. \tag{6.33}$$

Since the Jacobi field $J(\tau)$ can also be expressed as

$$J(\tau) = \eta(\tau)\int_{-T/2}^{\tau} \frac{d\tau'}{\eta^2(\tau')}, \tag{6.34}$$

we have at $\tau = T/2$

$$J(T/2) = \eta(T/2)\int_{-T/2}^{T/2} \eta^{-2}(\tau')d\tau'. \tag{6.35}$$

At large negative τ, we can neglect the difference between $\Omega(\tau)$ and Ω_0 and thus we have

$$\eta(\tau) \simeq \exp[\Omega_0(\tau + T/2)]. \tag{6.36}$$

Then we have

$$\eta(\tau) = \exp[\Omega_0 T/2]\eta_0(\tau), \tag{6.37}$$

where $\eta_0(\tau)$ is the solution with asymptotic behavior $\exp[\Omega_0\tau]$ at negative τ. We next note that since all the eigenvalues of $[-\partial^2/\partial\tau^2 + \Omega^2(\tau)]$ are positive, $\eta_0(\tau) \neq 0$ everywhere. At large positive τ, its asymptotic behavior is generally given by two exponentials $\exp[\pm\Omega_0\tau]$. Again, since the operator for the transverse fluctuations $[-\partial^2/\partial\tau^2 + \Omega^2(\tau)]$ does not have zero eigenvalues, the coefficient for the growing exponential $\exp(\Omega_0\tau)$ is not zero. We thus conclude that as T grows, $\eta(\tau)$ becomes exponentially large everywhere in the interval $[-T/2, T/2]$ except for the vicinity of $\tau = -T/2$, and that we can substitute $\eta(\tau)$ in Equation (6.35) by its asymptotic behavior, Equation (6.36). Then we have

$$J(T/2) = \frac{\eta(T/2)}{2\Omega_0}. \tag{6.38}$$

To proceed further, we define two more functions. Let $\tilde{\eta}(\tau)$ be a solution of the Jacobi equation that satisfies the following initial conditions at $\tau = T/2$ [due to the symmetry of $\Omega(\tau)$ this solution is just $\tilde{\eta}(\tau) = \eta(-\tau)$]:

$$\tilde{\eta}(T/2) = 1, \quad \dot{\tilde{\eta}}(T/2) = -\Omega_0. \tag{6.39}$$

Introducing the auxiliary function $\Theta(\tau)$ by

$$\eta(\tau) = \tilde{\eta}(\tau)\Theta(\tau) \tag{6.40}$$

and using the evident relations,

$$\eta(0) = \tilde{\eta}(0), \quad \dot{\eta}(0) = -\dot{\tilde{\eta}}(0), \tag{6.41}$$

we find

$$\Theta(\tau) = 1 + 2\eta(0)\dot{\eta}(0)\int_0^{\tau} \tilde{\eta}^{-2}(\tau')d\tau'. \tag{6.42}$$

At $\tau = T/2$ we can neglect the unity compared to the second term $\sim \exp(\Omega_0 T)$ and evaluate the integral in the same way as before. Now all the contribution to the integral at large T comes from the upper limit and we obtain

$$\eta(T/2) = \Theta(T/2) \simeq \frac{\eta(0)\dot{\eta}(0)}{\Omega_0}. \tag{6.43}$$

Using Equations (6.38) and (6.43), we arrive at the expression

$$\frac{J(T/2)}{J_0(T/2)} = \frac{\eta(0)\dot{\eta}(0)}{\Omega_0}\exp[-\Omega_0 T]. \tag{6.44}$$

We carry out one more transformation to make the final answer more convenient for practical use. Introducing the function

$$\Xi(\tau) = \dot{\eta}(\tau)\eta^{-1}(\tau), \tag{6.45}$$

which satisfies

$$\dot{\Xi}(\tau) + \Xi^2(\tau) = \Omega^2(\tau) \tag{6.46}$$

with the initial condition

$$\Xi(-T/2) = \Omega_0, \tag{6.47}$$

and noticing the relation

$$\eta(\tau) = \exp\left(\int_{-T/2}^{\tau} \Xi(\tau')d\tau'\right), \tag{6.48}$$

we get [see Equation (6.44)]

$$\frac{J(T/2)}{J_0(T/2)} = \frac{\Xi(0)}{\Omega_0}\exp\left[2\int_{-T/2}^{0}(\Xi(\tau)-\Omega_0)d\tau\right]. \tag{6.49}$$

The final result for the transversal pre-exponential factor is obtained by putting $T=\infty$ in the above integral as well as in the initial condition for $\Xi(\tau)$ in Equation (6.47).

Generalization to the case $N>2$ can be made rather straightforwardly. The transverse factor in Equation (6.26) is expressed in terms of the determinants of the Jacobi fields that satisfy the matrix equations analogous to Equation (2.112). The denominator $\mathbf{J}_0(\tau)$ is just a diagonal matrix and the analog of Equation (6.30) for $N>2$ is

$$\frac{\det\mathbf{J}(T/2)}{\det\mathbf{J}_0(T/2)} = 2^{N-1}\det\mathbf{\Omega}_0\det\mathbf{J}(T/2)\exp[-(\mathrm{Tr}\mathbf{\Omega}_0)T], \tag{6.50}$$

where $\mathbf{J}(\tau)$ is the solution of the initial value problem

$$\ddot{\mathbf{J}}(\tau) = \mathbf{\Omega}^2(\tau)\mathbf{J}(\tau) \tag{6.51}$$

with

$$\mathbf{J}(-T/2)=0,\quad \dot{\mathbf{J}}(-T/2)=1. \tag{6.52}$$

As before, we introduce the auxiliary matrix $\hat{\eta}(\tau)$, which is the solution of Equation (6.51) with the initial conditions

$$\hat{\eta}(-T/2)=\mathbf{1},\quad \dot{\hat{\eta}}(-T/2)=\Omega_0. \tag{6.53}$$

For the Jacobi field we obtain

$$\mathbf{J}(T/2) = \hat{\eta}(T/2)\left[\int_{-T/2}^{T/2}\hat{a}(\tau')d\tau'\right], \tag{6.54}$$

where the matrix $\hat{a}(\tau)$ is the solution of the first-order matrix equation

$$\hat{\eta}(\tau)\frac{d}{d\tau}\hat{a}(\tau)+2\frac{d}{d\tau}\hat{\eta}(\tau)\hat{a}(\tau)=0 \tag{6.55}$$

with

$$\hat{a}(-T/2)=\mathbf{1}. \tag{6.56}$$

As before, $\hat{\eta}(\tau)$ has the form $\exp(\mathbf{\Omega}_0 T/2)$ multiplied by a matrix solution of Equation (6.51) decaying at negative τ as $\exp(\mathbf{\Omega}_0\tau)$. Under the assumption that all the eigenvalues of the operator $[\partial^2/\partial\tau\mathbf{1}+\mathbf{\Omega}^2]$ are positive, the determinant of this solution is not zero and the inverse matrix $\hat{\eta}^{-1}$ exists. Then it can be shown that the elements of $\hat{a}(\tau)$ are exponentially small everywhere except at large negative τ where the asymptotic form

$$\hat{a}_{ij}\simeq\delta_{ij}\exp[-2\mathbf{\Omega}_i(\tau+T/2)] \tag{6.57}$$

can be used. Then the integration in Equation (6.54) gives

$$\det\mathbf{J}(T/2)=\frac{\det\hat{\eta}(T/2)}{2^{N-1}\det\mathbf{\Omega}_0}. \tag{6.58}$$

In turn, the analog of Equation (6.43) is

$$\det \hat{\eta}(T/2) = \frac{\det \hat{\eta}(0) \det \dot{\hat{\eta}}(0)}{\det \boldsymbol{\Omega}_0}. \tag{6.59}$$

The easiest way to prove the last relation is to divide $[-T/2, T/2]$ into a number of intervals $[-\epsilon(k-1), \epsilon k]$ with $k = 1 - T/\epsilon, \ldots, T/\epsilon$ and to approximate $\boldsymbol{\Omega}^2(\tau)$ by a steplike matrix function $\tilde{\boldsymbol{\Omega}}^2(\tau)$ that takes constant values within each interval of the length ϵ. Equation (6.59) for $\tilde{\boldsymbol{\Omega}}(\tau)$ can be obtained in the limit $T \to \infty$ by direct but quite tedious algebra. The true matrix function $\Omega(\tau)$ then can be considered as a limit of infinitesimally small step size. Finally, we introduce the matrix $\Xi(\tau) = [(d/d\tau)\hat{\eta}(\tau)]\hat{\eta}^{-1}(\tau)$, which satisfies the matrix analog of Equation (6.46)

$$\dot{\Xi} + \Xi^2 = \boldsymbol{\Omega}^2(\tau) \tag{6.60}$$

and

$$\Xi(-T/2) = \boldsymbol{\Omega}_0. \tag{6.61}$$

Taking the derivative of $\det \hat{\eta}$, we find

$$\frac{d}{d\tau} \det \hat{\eta} = \det \hat{\eta} \sum_{ij} \dot{\hat{\eta}}_{ij} \hat{\eta}_{ij}^{-1} = \det \hat{\eta} \cdot [\mathrm{Tr}\Xi], \tag{6.62}$$

or in view of the initial conditions for $\hat{\eta}$ and Ξ

$$\det \hat{\eta}(\tau) = \exp\left(\int_{-T/2}^{\tau} [\mathrm{Tr}\Xi(\tau')] d\tau' \right). \tag{6.63}$$

In the limit $T \to \infty$, Equations (6.26), (6.50), (6.58), (6.59), and (6.63) give the transverse pre-exponential factor for multidimensional case as

$$\sqrt{\frac{\det \boldsymbol{\Omega}_0}{\det \Xi(0)}} \exp\left[\int_{-\infty}^{0} \mathrm{Tr}\,(\boldsymbol{\Omega}_0 - \Xi(\tau))\, d\tau \right]. \tag{6.64}$$

Equations (2.135) and (6.64) solve the problem of calculating Δ_0. Of course, one still has to find the instanton trajectory and to solve the system of the differential equations for the matrix Ξ. The latter requires the Hessian of the potential in terms of the transverse local coordinates together with the corresponding curvatures.

6.1.2 WKB Approach in Cartesian Coordinates

The results obtained in the previous subsection of instanton theory can be reproduced in a much easier way with use of the WKB approximation. This is quite helpful to understand the physical meaning of the various quantities and the procedures used. In the same way as we did in the one-dimensional case in Section 2.5.2, we do not need to construct the complex valued Lagrange manifold [7,15,30,37], which constitutes the main obstacle of the conventional WKB theory. Without the energy term, the Hamilton-Jacobi equation can be easily solved and generalization for an arbitrary number of coordinates is straightforward. For the sake of completeness and smooth introduction to the new method, we first outline the derivation of the transverse pre-exponential factor previously given in Reference [53]. We use the local coordinates $(S, \{\xi\})$ used before and choose the symmetric parametrization of the instanton $\mathbf{q}_0(S)$, $S \in [-S_m, S_m]$, where $\mathbf{q}_0(\pm S_m)$ represent the potential minima. In these coordinates the multidimensional Herring formula reads

$$\Delta = -\frac{2\int d\xi_1 \cdots d\xi_{N-1}\Psi \frac{\partial\Psi}{\partial S}|_{S=0}}{\int d\mathbf{q}\Psi^2(\mathbf{q})} \tag{6.65}$$

and within the accuracy of the semiclassical approximation the integrations are to be carried out in the Gaussian approximation. We first consider the main exponential factor W_0, which is the solution of the Hamilton-Jacobi equation

$$\frac{1}{2(1+\sum_\alpha C_\alpha\xi_\alpha)^2}\left(\frac{\partial W_0}{\partial S}\right)^2 + \frac{1}{2}\sum_\alpha\left(\frac{\partial W_0}{\partial \xi_\alpha}\right)^2 = V_0(S) + \sum_\alpha a_\alpha\xi_\alpha + \frac{1}{2}\sum_{\alpha\beta} b_{0\alpha\beta}\xi_\alpha\xi_\beta + o(\xi^2). \tag{6.66}$$

We expand $1/(1+\sum_\alpha C_\alpha\xi_\alpha)$ and sequentially equate the terms according to the orders ξ^0, ξ^1, ξ^2. This gives $W_0(S, \{\xi\})$ in the form

$$W_0(S, \{\xi_\alpha\}) = \int_{-S_m}^{S} p_0(S')dS' + \frac{1}{2}\sum_{\alpha\beta}\Xi_{\alpha\beta}(S)\xi_\alpha\xi_\beta, \tag{6.67}$$

where $p_0(S) = \sqrt{2V_0(S)}$ is the absolute value of the momentum on the instanton trajectory and $\Xi_{\alpha\beta}(S)$ satisfies the equation [see Equation (6.27)]

$$3p_0^2 C_\alpha C_\beta + p_0\frac{d\Xi_{\alpha\beta}}{dS} + \Xi_{\alpha\beta}^2 = b_{\alpha\beta}. \tag{6.68}$$

Since along the instanton $p_0(S)\partial/\partial S = d/d\tau$, this equation coincides with Equation (6.60) of the previous subsection. In turn, according to the initial condition Equation (6.61) in the vicinity of the potential minimum $(S = -S_m, \xi_\alpha = 0)$, we have

$$\begin{aligned} W_0 &= \int_{-S_m}^{S} \omega_N(S'+S_m)dS' + \frac{1}{2}\sum_{\alpha\beta}\Xi_{\alpha\beta}(-S_m)\xi_\alpha\xi_\beta \\ &= \frac{1}{2}\omega_N(S+S_m)^2 + \frac{1}{2}\sum_{\alpha=1}^{N-1}\omega_\alpha\xi_\alpha^2 \end{aligned} \tag{6.69}$$

and that $\exp(-W_0)$ behaves as the wave function of the ground state of an N-dimensional harmonic oscillator with the energy $E = (\hbar/2)\sum_\alpha^N \omega_\alpha$.

The integration of the transport equation is trivial and for the present purpose we only need the solution on the instanton trajectory

$$W_1(S) = \frac{1}{2}\ln p_0(S) - \frac{1}{2}\int^S \frac{\omega_N}{p_0(S')}dS' - \frac{1}{2}\int_{-S_m}^{S}\frac{dS'}{p_0(S')}\left[\sum_{\alpha=1}^{N-1}(\omega_\alpha - \Xi_{\alpha\alpha})\right]. \tag{6.70}$$

We note that the only difference from the one-dimensional case comes from the last term. Inserting the expression of the wave function

$$\Psi = \exp\left(-\frac{W_0}{\hbar} - W_1\right) \tag{6.71}$$

with Equations (6.69) and (6.70) into Equation (6.65), we find the splitting as usual in the form

$$\Delta_0 = B\exp(-S_0) \tag{6.72}$$

with

$$B = \sqrt{\frac{4\omega_N}{\pi}}\, p_0(0)\exp\left(\int_{-\infty}^{0} d\tau\left[\omega_N - \frac{\partial p_0(S)}{\partial S}\right]\right)$$
$$\times\sqrt{\frac{\det \boldsymbol{\Omega}_0}{\det \Xi(0)}}\exp\left(\int_{-\infty}^{0} \mathrm{Tr}(\boldsymbol{\Omega}_0 - \Xi(\tau))d\tau\right). \tag{6.73}$$

We can see that the transverse pre-exponential factor is the same as in Equation (6.64) and the present WKB approach is equivalent to the instanton approach.

The calculations of $\Xi_{\alpha\beta}$ in terms of the local coordinates are not convenient, but this can now be avoided and an invariant formulation can be performed. The matrix $\Xi_{\alpha\beta} = \partial^2 W_0/\partial\xi_\alpha\partial\xi_\beta$ [see Equation (6.67)] describes the Gaussian shape of the wave function in the plane orthogonal to the instanton. It is actually more advantageous to use the analogous "full" N by N symmetric matrix

$$A_{ij} = \frac{\partial^2 W_0}{\partial q_i \partial q_j}. \tag{6.74}$$

Along the instanton this matrix satisfies a more handy equation

$$\frac{d}{d\tau}A_{ij} + (\mathbf{A}^2)_{ij} = \frac{\partial^2 V(\mathbf{q})}{\partial q_i \partial q_j}|_{\mathbf{q}=\mathbf{q}_0(\tau)}, \tag{6.75}$$

which can be obtained directly by differentiating the Hamilton-Jacobi equation. An analogous more general equation is derived in the next subsection. The initial condition of $\mathbf{A}$ follows from the fact that in the vicinity of the potential minimum $\exp(-W_0)$ coincides with the ground-state wave function of the harmonic oscillator and is given by

$$\mathbf{A}^2_{ij}|_{\tau\to-\infty} = \frac{\partial^2 V}{\partial q_i \partial q_j}|_{\mathbf{q}=\mathbf{q}_m}. \tag{6.76}$$

Using the definition of the local coordinates S and $\{\xi_\alpha\}$, we further find that

$$\Xi_{\alpha\beta} = \sum_{ij=1}^{N} A_{ij}\tau_{i\alpha}\tau_{j\beta} \tag{6.77}$$

and

$$A_{ij} = \tau_{iN}\tau_{jN}p_0' + \sum_{\alpha\beta}\tau_{i\alpha}\tau_{j\beta}\Xi_{\alpha\beta} - \sum_{\alpha} p_0 C_{0\alpha}(\tau_{iN}\tau_{j\alpha} + \tau_{i\alpha}\tau_{jN}). \tag{6.78}$$

These relations enable us to rewrite Equation (6.73) totally in terms of A_{ij}. First, it is easy to see that $\omega_N \Pi_\alpha \omega_\alpha = \det \mathbf{A}(-\infty)$ and that due to Equation (6.78) the two integral terms in the pre-exponential factor can be combined together as

$$\int d\tau \left(\omega_N - \frac{dp_0(S)}{dS}\right) + \int d\tau \sum_\alpha (\omega_\alpha - \Xi_{\alpha\alpha}) = \int d\tau \mathrm{Tr}(\mathbf{A}(-\infty) - \mathbf{A}(\tau)). \quad (6.79)$$

Noting that the determinant of $\Xi_{\alpha\beta}$ comes from the integration over $N-1$ transverse coordinates ξ_α, using the relation between $\Xi_{\alpha\beta}$ and A_{ij}, and making the coordinate transformation of Equation (6.17), we obtain (see Appendix B)

$$\begin{aligned}
\frac{1}{\sqrt{\det \Xi}} &= \frac{1}{\sqrt{\pi}^{N-1}} \int d\xi_1 \cdots d\xi_{N-1} \exp\left(-\sum_{ij\alpha\beta} A_{ij}\tau_{i\alpha}\tau_{j\beta}\xi_\alpha\xi_\beta\right) \\
&= \frac{1}{\sqrt{\pi}^{N-1}} \int dq_1 \cdots dq_N \delta\left(\sum_i \tau_{iN} q_i\right) \exp\left(-\sum_{ij} A_{ij} q_i q_j\right) \\
&= \frac{1}{\sqrt{\det \mathbf{A} \left(\sum_{ij} \tau_{iN} A^{-1}_{ij} \tau_{jN}\right)}} \\
&= \frac{|\mathbf{p}|}{\sqrt{\det \mathbf{A} \left(\sum_{ij} p_i A^{-1}_{ij} p_j\right)}}, \qquad (6.80)
\end{aligned}$$

where $\mathbf{p}(\tau) = \dot{\mathbf{q}}_0(\tau)$. Putting together Equations (6.73), (6.79), and (6.80), we finally obtain the tunneling splitting in Cartesian coordinates as

$$\begin{aligned}
\Delta_0 &= \sqrt{\frac{4 \det \mathbf{A}(-\infty)}{\pi \det \mathbf{A}(0)}} \frac{\mathbf{p}^2(0)}{\sqrt{\sum_{ij} p_i(0) A^{-1}_{ij} p_j(0)}} \\
&\quad \times \exp\left(\int_{-\infty}^{0} d\tau \left[\mathrm{Tr}\left(\mathbf{A}(-\infty) - \mathbf{A}(\tau)\right)\right]\right) \exp(-S_0). \qquad (6.81)
\end{aligned}$$

Thus, once the instanton trajectory is found, the calculations of Δ_0 are straightforward. In fact, within the framework of this type of WKB approach the local coordinates are not necessary at all.

6.1.3 WKB Approach in the Case of General Hamiltonian in Curved Space

Finally in this subsection, the most general formulation is provided for a general form of kinetic-energy operator without relying on any local coordinates. The final expression of tunneling splitting can be applied to any practical systems.

Let $\mathbf{q} = (q^1, \ldots, q^N)$ be coordinates on a manifold M with Riemannian metric $q_{ij}(\mathbf{q})$ and let $V(\mathbf{q})$ be a potential function that is symmetric and has two equivalent minima located at $\mathbf{q}_m$ and $\tilde{\mathbf{q}}_m = C\mathbf{q}_m$, where C represents the symmetry operation:

$C : M \to M$ with $C^{-1} = C$. The Hamiltonian operator is given by

$$\hat{H} = -\frac{1}{2\sqrt{g}} \sum_{ij} \frac{\partial}{\partial q^i} \left(\sqrt{g} g^{ij} \frac{\partial}{\partial q^j} \right) + V(\mathbf{q}), \tag{6.82}$$

where, as usual, $g = \det(g_{ij})$ and g^{ij} is the inverse of g_{ij}. Hereafter the energy is measured from the potential minima, namely, $V(\pm\mathbf{q}_m) = 0$. We also introduce an $(N-1)$-dimensional "dividing" surface Σ as a solution of

$$f(\mathbf{q}) = 0. \tag{6.83}$$

The dividing surface Σ has a simple meaning of the location of a potential barrier that separates two potential wells. The tunneling splitting is now given by

$$\Delta_0 = \frac{2 \int \sqrt{g} d\mathbf{q} \delta(f) \Psi \sum_{ij} g^{ij} \frac{\partial \Psi}{\partial q^i} \frac{\partial f}{\partial q^j}}{\int \sqrt{g} d\mathbf{q} \Psi^2}, \tag{6.84}$$

which is the analog of the Herring formula for an arbitrary metric and is derived in Appendix C. Our purpose here is to find the wave function $\Psi(\mathbf{q})$ that is localized near the potential minimum $\mathbf{q}_m$ and has the form, as usual,

$$\Psi = \exp\left(-\frac{W_0}{\hbar} - W_1 \right). \tag{6.85}$$

Then two integrals in Equation (6.84) can be carried out in the Gaussian approximation. For the present case, the Hamilton-Jacobi equation and the transport (continuity) equation read

$$H\left(\mathbf{q}, \frac{\partial W_0}{\partial \mathbf{q}} \right) = 0 \tag{6.86}$$

and

$$\sum_{ij} g^{ij}(q) \frac{\partial W_0}{\partial q^i} \frac{\partial W_1}{\partial q^j} - \frac{1}{2} \sum_{ij} g^{ij}(q) \frac{\partial^2 W_0}{\partial q^i \partial q^j} + \frac{E}{\hbar} - \frac{1}{2} \sum_i \Lambda^i \frac{\partial W_0}{\partial q^i} = 0, \tag{6.87}$$

where $H(\mathbf{q}, \mathbf{p})$ is the classical Hamiltonian

$$H(\mathbf{q}, \mathbf{p}) = \frac{1}{2} \sum_{ij} g^{ij}(\mathbf{q}) p_i p_j - V(\mathbf{q}), \tag{6.88}$$

and the last term in Equation (6.87) appears due to the $\mathbf{q}$-dependence of the metric

$$\Lambda^i(\mathbf{q}) = \frac{1}{\sqrt{g}} \sum_j \frac{\partial}{\partial q^j} (\sqrt{g} g^{ij}). \tag{6.89}$$

To solve Equation (6.86) we use the method of characteristics and consider a family of classical trajectories $[\mathbf{q}(\beta, \tau), \mathbf{p}(\beta, \tau)]$ where β is an $(N-1)$-dimensional parameter and τ is the time running the infinite time interval. According to Equation (6.86) we put the energy to be zero and impose the initial conditions $\mathbf{q}(\beta, \tau = -\infty) = \mathbf{q}_m$, $\mathbf{p}(\beta, \tau = -\infty) = 0$, which defines the N-dimensional Lagrange manifold. We further note that the instanton trajectory $\mathbf{q}_0(\tau)$ belongs to this family $\mathbf{q}_0(\tau) = \mathbf{q}(\beta_0, \tau)$ and we choose such a parametrization that $C\mathbf{q}(\beta_0, \tau) \equiv \mathbf{q}^C(\beta_0, \tau) = \mathbf{q}(\beta_0, -\tau)$. Once the family of characteristics is fixed, $W_0(\mathbf{q})$ is given by the integral of one-form $\sum_i p_i dq^i$. At $\mathbf{q}$ close to the potential minimum $\mathbf{q}_m$, the momentum $\mathbf{p}(\mathbf{q})$ is linear in the deviation $(\mathbf{q} - \mathbf{q}_m)$ and $W_0(\mathbf{q})$ is quadratic

$$W_0 = \frac{1}{2}(\mathbf{q} - \mathbf{q}_m)^T \mathbf{A}_m (\mathbf{q} - \mathbf{q}_m) + o\left((\mathbf{q} - \mathbf{q}_m)^2\right). \tag{6.90}$$

Thus, the imposed initial conditions for the characteristics are equivalent to the requirement that the semiclassical solution $\exp(-W_0(\mathbf{q})/\hbar)$ at the potential minimum coincides with the wave function of the ground state of harmonic oscillator. Inserting Equation (6.90) into the Hamilton-Jacobi Equation (6.86), we come to the relation

$$\mathbf{A}_m \mathbf{g}_m \mathbf{A}_m = V_{\mathbf{qq}}(\mathbf{q}_m), \tag{6.91}$$

where $V_{\mathbf{qq}}$ is the matrix of the second derivatives $\partial^2 V/\partial q^i \partial q^j$ and $\mathbf{g}_m$ is the compact notation for $g^{ij}(\mathbf{q}_m)$. The matrix $\mathbf{A}_m$ is all we need to evaluate the normalization integral $\int d\mathbf{q}\Psi^2(\mathbf{q})$. For the integral in the numerator of the Herring formula we have to know the behavior of the semiclassical solution on the dividing surface Σ.

Let $\mathbf{q}_\Sigma = \mathbf{q}_0(0)$ be the point where $\mathbf{q}_0(\tau)$ crosses Σ. $W_0^C(\mathbf{q}) \equiv W_0(C\mathbf{q})$ is also a solution of the Hamilton-Jacobi equation, which corresponds to the wave function localized near $C\mathbf{q}_m$. On the dividing surface Σ these two solutions coincide and if $u_k (k = 1, 2, \ldots, N-1)$ are the coordinates on Σ, we have

$$\sum_i \frac{\partial W_0}{\partial q^i}\frac{\partial q^i}{\partial u_k} = \sum_i \frac{\partial W_0^C}{\partial q^i}\frac{\partial q^i}{\partial u_k}. \tag{6.92}$$

On the other hand, on the instanton trajectory we have

$$\frac{\partial W_0\left(\mathbf{q}_0(\tau)\right)}{\partial q^i} = \sum_j g_{ij}\left(\mathbf{q}_0(\tau)\right)\dot{q}_0^j(\tau) = -\sum_j g_{ij}\left(\mathbf{q}_0^C(-\tau)\right)\dot{q}_0^{Cj}(-\tau)$$

$$= -\frac{\partial W_0^C\left(\mathbf{q}_0(\tau)\right)}{\partial q^i}. \tag{6.93}$$

Comparing Equation (6.92) and Equation (6.93), we find

$$\sum_i \frac{\partial W_0}{\partial q^i}\frac{\partial q^i}{\partial u_k}\Big|_{\mathbf{q}=\mathbf{q}_\Sigma} = 0, \tag{6.94}$$

which means that $\mathbf{q}_\Sigma$ is the stationary point for the integral over Σ. This is equivalent to the following relation between the derivatives of $f(\mathbf{q})$ and the momentum $\mathbf{p}$ at the stationary point:

$$\frac{\partial W_0(\mathbf{q}_\Sigma)}{\partial q^i} = p_i(\mathbf{q}_\Sigma) = \alpha\frac{\partial f_\Sigma}{\partial q^i}, \tag{6.95}$$

where α is some constant. Next, we need to find the matrix of the second derivatives

$$\tilde{A}_{ij} \equiv \frac{\partial^2 W_0}{\partial q^i \partial q^j} = \frac{\partial p_i}{\partial q^j} \tag{6.96}$$

at the stationary point $\mathbf{q}_\Sigma$. Let us consider the matrix $\tilde{\mathbf{A}}(\tau)$ as a function of time along the instanton. From the classical Hamilton equations for the characteristics, we obtain the first-order differential equation (see Appendix D)

$$\dot{\tilde{\mathbf{A}}} = -H_{\mathbf{qq}} - H_{\mathbf{qp}}\tilde{\mathbf{A}} - \tilde{\mathbf{A}}H_{\mathbf{pq}} - \tilde{\mathbf{A}}H_{\mathbf{pp}}\tilde{\mathbf{A}}. \tag{6.97}$$

Here $H_{\mathbf{qq}}, H_{\mathbf{qp}}, \ldots$ are the matrices of the second derivatives $\frac{\partial^2 H}{\partial q^i \partial q^j}, \frac{\partial^2 H}{\partial q^i \partial p_j}, \ldots$. Equation (6.97) must be supplemented by the initial condition

$$\tilde{\mathbf{A}}(-\infty) = \mathbf{A}_m, \tag{6.98}$$

which completely determines $\tilde{\mathbf{A}}(\tau)$. In particular, $\tilde{\mathbf{A}}(\tau = 0)$ is the value of the matrix at the stationary point $\mathbf{q}_\Sigma$.

Now, we go back to the evaluation of the Herring formula. The normalization integral is trivial. Expanding the argument of the exponent and δ-function, we rewrite Δ_0 as

$$\Delta_0 = 2\sqrt{\frac{g_\Sigma \det \mathbf{A}_m}{g_m \pi^N}} \left(\sum_{ij} g^{ij} \frac{\partial W_0}{\partial q^i} \frac{\partial f}{\partial q^j} \right)_\Sigma \exp(-2W_{0\Sigma} - 2W_{1\Sigma})$$
$$\times \int d\mathbf{q}\,\delta\left(\sum_i \frac{\partial f_\Sigma}{\partial q^i} q^i + \frac{1}{2} \sum_{ij} \frac{\partial^2 f_\Sigma}{\partial q^i \partial q^j} q^i q^j \right) \exp\left[-2\mathbf{p}_\Sigma^T \mathbf{q} - \mathbf{q}^T \tilde{\mathbf{A}}_\Sigma \mathbf{q} \right], \tag{6.99}$$

where the labels m and Σ indicate that the corresponding quantities are taken at $\mathbf{q}_m$ and $\mathbf{q}_\Sigma$, respectively. The principal exponential factor $2W_{0\Sigma}$ is the action S_0 along the instanton and $W_{1\Sigma}$ is given below. Performing the integration and taking into account Equation (6.95), we obtain

$$\Delta_0 = \sqrt{\frac{4g_\Sigma \det \mathbf{A}_m}{\pi g_0 \det \mathbf{A}_\Sigma}} \frac{(\mathbf{p}^T \mathbf{g} \mathbf{p})_\Sigma}{\sqrt{(\mathbf{p}^T \mathbf{A}^{-1} \mathbf{p})_\Sigma}} \exp(-2W_{0\Sigma} - 2W_{1\Sigma}), \tag{6.100}$$

where

$$\mathbf{A}_\Sigma \equiv \tilde{\mathbf{A}}_\Sigma + \delta\tilde{\mathbf{A}}_\Sigma \tag{6.101}$$

with

$$(\delta\tilde{\mathbf{A}}_\Sigma)_{ij} = -\alpha \frac{\partial^2 f_\Sigma}{\partial q^i \partial q^j}. \tag{6.102}$$

This is not the final result yet, since the dividing surface Σ is an auxiliary object in the theory and $\partial^2 f_\Sigma / \partial q^i \partial q^j$ must be excluded from the expression. The term $\delta\tilde{\mathbf{A}}_\Sigma$ appears due to the curvature of the dividing surface. It accounts for the curvature of the space itself and must be related with the derivative of the metric tensor. Equation (6.102) also shows that $\delta\tilde{\mathbf{A}}_\Sigma$ is proportional to the momentum $\mathbf{p}_\Sigma$. Then the explicit form of $\delta\tilde{\mathbf{A}}_\Sigma$ is uniquely established by the condition that $\mathbf{A}_\Sigma$ must transform as a tensor under arbitrary nonlinear coordinate transformation in order to have an invariant expression for the tunneling splitting. This gives

$$(\delta\tilde{\mathbf{A}}_\Sigma)_{ij} = -\sum_k p_k(\mathbf{q}_\Sigma)\Gamma^k_{ij}(\mathbf{q}_\Sigma), \tag{6.103}$$

where

$$\Gamma^k_{ij} = \frac{1}{2}\sum_s g^{ks}\left(\frac{\partial g_{is}}{\partial q^j} + \frac{\partial g_{js}}{\partial q^i} - \frac{\partial g_{ij}}{\partial q^s}\right) \tag{6.104}$$

are the Christoffel symbols. [For the flat space, Equation (6.103) can be obtained directly from the results derived in the Cartesian coordinates.]

The value of $W_{1\Sigma}$ is found from the transport (continuity) equation, Equation (6.87). Since $\sum_{ij} g^{ij}[\partial W_0/\partial q^i]\partial/\partial q^j$ is nothing but the time derivative along the instanton $d/d\tau$, the direct integration gives

$$2W_{1\Sigma} = \int_{-\infty}^{0} d\tau \left[\mathrm{Tr}(\tilde{\mathbf{A}}(\tau) - \mathbf{A}_m) + \mathbf{p}^T\mathbf{\Lambda}\right], \tag{6.105}$$

where $\mathrm{Tr}(\tilde{\mathbf{A}}) \equiv \sum_i \tilde{A}^i_i = \sum_{ij} g^{ij}\tilde{A}_{ji}$ and we also took into account that the following relation holds for the ground state:

$$\frac{E}{\hbar} = \frac{1}{2}\mathrm{Tr}(\mathbf{A}_m) + O(\hbar). \tag{6.106}$$

Equations (6.99) and (6.105) are the final result. It gives the canonically invariant expression of the tunneling splitting, and actual computations can be carried out in any convenient system of coordinates. Once the instanton trajectory is found, the only thing remaining is to solve the system of the first-order differential equations, Equation (6.97), which does not constitute any numerical problem. One useful remark can be made about the form of kinetic-energy operator. In general, it may differ from the canonical form in Equation (6.82) by first-order derivative terms, which leads only to evident modification of Λ^i in the transport equation, Equation (6.87).

6.2 HOW TO FIND INSTANTON TRAJECTORY

In order to implement the theory given in the previous section, we must be able to find the instanton trajectory $\mathbf{q}_0(\tau)$ in a multidimensional space. There are two problems. First of all, it would be hopeless to try to find the instanton trajectory in high-dimensional space by simply "shooting" classical trajectories. Some approximate forms of the instanton such as a combination of minimum energy path and straight line path are often used in the literature. Such approximations, however, affect the accuracy of the principal exponential factor S_0, and can easily and erroneously change the orders of magnitude of the final results. Secondly, the accuracy of the results strongly depends on the quality of potential energy surface employed. Thanks to recent advances in electronic structure theory and computational technologies, the electronic energy at a given molecular configuration can be calculated with a reasonable computational cost. However, construction of accurate global full-dimensional potential energy surface is not straightforward and one has to find an effective way to incorporate the *ab initio* data into dynamics simulations. The first problem is discussed in this subsection and the second one—namely, the efficient way of using highly accurate *ab initio* data—is presented in the next subsection.

The practical method to find the instanton trajectory is not to solve the classical equations of motion but to directly find the instanton *path* by minimizing the Euclidean action. Let us consider the Lagrangian with the inverted potential

$$L(\dot{\mathbf{q}}, \mathbf{q}) = \frac{1}{2} \sum_{ij} g_{ij}(\mathbf{q})\dot{q}_i\dot{q}_j + V(\mathbf{q}), \tag{6.107}$$

where $g_{ij}(\mathbf{q})$ is the inverse of $g^{ij}(\mathbf{q})$. Let $\mathbf{q}_m$, $\tilde{\mathbf{q}}_m$ be the positions of the two potential minima $V(\mathbf{q}_m) = V(\tilde{\mathbf{q}}_m) = 0$. By definition, the instanton $q_0(\tau)$ with $\tau \in [-\infty, \infty]$ is the classical trajectory satisfying the boundary conditions

$$\mathbf{q}_0(\tau = -\infty) = \mathbf{q}_m, \quad \mathbf{q}_0(\tau = \infty) = \tilde{\mathbf{q}}_m. \tag{6.108}$$

In a multidimensional case, however, it would be hopeless, as mentioned above, to find the right instanton trajectory that satisfies the boundary conditions, Equation (6.108), at both ends by simply shooting classical trajectories. The practical method is to use the variational principle to minimize the action,

$$S_0[\mathbf{q}_0(\tau)] = \int_{-\infty}^{\infty} L[\dot{\mathbf{q}}(\tau), \mathbf{q}(\tau)]d\tau. \tag{6.109}$$

Instead of using the time variable τ, it is more convenient to introduce some other parametrization. Without loss of generality, we introduce a new parameter z that spans the interval $[-1, 1]$ and call $\mathbf{q}_0(z)$ a *path* to distinguish from *trajectory* $\mathbf{q}_0(\tau)$. There is naturally one-to-one correspondence between path and classical trajectory and the correspondence can be established by the energy conservation, which defines the "velocity" of the parameter z as

$$\dot{z}(z) = \sqrt{\frac{2V(\mathbf{q}_0(z))}{\sum_{ij} g_{ij}(\mathbf{q}_0(z)) \frac{dq_0^i(z)}{dz}\frac{dq_0^j(z)}{dz}}}, \tag{6.110}$$

where the dot indicates the derivative with respect to time τ. The proposed method is to minimize the classical action in the space of paths in analytical form:

$$q_0^i(z, \{C^{in}\}) = \frac{\tilde{q}_m^i + q_m^i}{2} + \frac{\tilde{q}_m^i - q_m^i}{2} z + \sum_{n=1}^{N_b} C^{in}\phi_n(z), \tag{6.111}$$

where $\phi_n(z)$ is a certain set of smooth basis functions under the condition $\phi_n(z = \pm 1) = 0$. The first two terms of Equation (6.111) represent a straight line connecting the two potential minima, and the purpose here is to find the best coefficients

$\{C^{in}\}$. In the actual calculations we use the basis

$$\phi_n(z) = (1 - z^2)P_n(z), \tag{6.112}$$

where $P_n(z)$ are the Legendre polynomials. It turns out that quite a small number of N_b of the basis functions enables us to attain high accuracy. The coefficients $\{C^{in}\}$ can be determined by the following iterative procedure:

1. Let $\{C^{in}\}$ be initial guess of the coefficients that gives the approximate form of Equation (6.111) of the instanton *path* $\mathbf{q}_0(z, \{C^{in}\})$. Integrating $dz/d\tau = \dot{z}(z)$ given by Equation (6.110), we can obtain the explicit time-dependence $z(\tau)$ and define the approximate instanton *trajectory* as $\mathbf{q}_0(\tau, \{C^{in}\}) \equiv \mathbf{q}_0[z(\tau), \{C^{in}\}]$. If the path $\mathbf{q}_0(z)$ is the correct one, the trajectory $\mathbf{q}_0[z(\tau)]$ would be the true instanton. Generally speaking, however, this is not the case and we have to improve it.
2. The main idea of improvement is to find a direction in the space of trajectories along which the classical action, Equation (6.109), decreases. In order to do so, we use $z(\tau)$ to define the τ-dependent functions $\phi_n[z(\tau)]$ that form the basis of expansion in the vicinity of $\mathbf{q}_0[z(\tau)]$. Namely, we look for a better instanton trajectory in the form

$$q_0^i(\tau) = q_0^i(z(\tau), \{C^{in}\}) + \sum_{n=1}^{N_b} \delta C^{in}\phi_n[z(\tau)]. \tag{6.113}$$

   Inserting Equation (6.113) into Equation (6.109), we obtain the classical action $S_0(\{C^{in}\})$ as a function of the coefficients $\{C^{in}\}$. Minimization of this function gives a new *trajectory* and accomplishes one step of iteration. The shape of the obtained *path* is again given by Equation (6.111) with the modified coefficients $\{C^{in} + \delta C^{in}\}$.
3. The iteration continues until we obtain the converged results. The initial guess of the instanton path can, of course, be taken in various ways, but it naturally affects the number of iterations required. Note, however, that the minimization of the action functional can be realized in a rather small subspace of paths formed by $\phi_n(z)$. The dimension of this subspace is $N \times N_b$, where N is the number of coordinates and N_b is the number of basis functions $\phi_n(z)$ that do not change in the course of iteration. The typical number of N_b is 10–30, as is seen in the actual applications to polyatomic molecules presented in the next chapter.

It should be noted that the explicit time-dependence $z(\tau)$ does not have to be determined in the course of iteration. Instead, since $\dot{z}(z)$ is given analytically, all the calculations can be carried out in terms of the parameter z; namely, we do not need the time τ. For example, the action functional is given by

$$S_0 = \int_{-1}^{1} dz\dot{z}(z)\left(\frac{1}{2}\sum_{kl} g_{kl}\left(\mathbf{q}_0(z)\right) q_0'^k(z) q_0'^l(z)\right) + \int_{-1}^{1} \frac{dz}{\dot{z}(z)} V\left(\mathbf{q}_0(z)\right), \tag{6.114}$$

where the prime stands for the derivative with respect to the parameter z and $\dot{z}(z)$ is the function of z determined by Equation (6.110). In order to minimize the action S_0 we assume the variation of the path in the form of Equation (6.113) and expand S_0 up to the second order with respect to δC^{in} as

$$S_0(\{\delta C^{in}\}) = S_0 + \delta(C^{in})^T \frac{\partial S_0}{\partial(\delta C^{in})} + \frac{1}{2}\delta(C^{in})^T \frac{\partial^2 S_0}{\partial(\delta C^{in})\partial(\delta C^{in})}\delta C^{in}. \quad (6.115)$$

This minimization gives a set of new coefficients and the improved instanton trajectory. Equation (6.115) requires gradient and Hessian of the potential function. For instance, the first derivative $\partial S_0/\partial\delta C^{in}$ reads

$$\frac{\partial S_0}{\partial\delta C^{in}} = \int_{-1}^{1} dz\dot{z}(z)\left(\frac{1}{2}\sum_{kl}\frac{\partial g_{kl}(\mathbf{q}_0(z))}{\partial q^i} q_0'^k(z) q_0'^l(z)\phi_n(z)\right.$$
$$\left. + \sum_k g_{ki}(\mathbf{q}_0(z)) q_0'^k(z)\phi_n'(z)\right) + \int_{-1}^{1}\frac{dz}{\dot{z}(z)}\frac{\partial V(\mathbf{q}_0(z))}{\partial q^i}\phi_n(z). \quad (6.116)$$

The second derivative $\partial^2 S_0/\partial\delta C^{in}\partial\delta C^{jm}$ is given by an integral akin to the above equation that includes the Hessian matrix $\partial^2 V[\mathbf{q}_0(z)]/\partial q^i \partial q^j$.

The Fortran source code to calculate instanton trajectory is provided in Appendix E. The reader may directly apply it to his/her own problem, although some standard subroutines should be prepared. One simple numerical example is provided with use of a model potential.

Since the quantum chemical computations of high accuracy requires a huge CPU time, we need some technique to save time. On a given approximate instanton path $\mathbf{q}_0(z)$ we choose N_r values of parameters $\{z_n\}_{n=1,2,\ldots,N_r}$ and define the corresponding set of N_r reference configurations $\{\mathbf{q}_0(z_n)\}$. Using the values of $V[\mathbf{q}_0(z_n)]$, $\partial V[\mathbf{q}_0(z_n)]/\partial q^i$, and $\partial^2 V[\mathbf{q}_0(z_n)]/\partial q^i \partial q^j$ at these reference points, we find the values at any intermediate z by piecewise smooth cubic interpolation procedure. Figure 6.1 illustrates this one-dimensional interpolation for the potential as well as the second derivatives. Figure 6.1(a) shows the potential function for the three kinds of quantum chemical *ab initio* methods used in the case of malonaldehyde. The symbols represent the *ab initio* calculations and the lines show the interpolation in the whole interval $z \in [-1, 1]$. The number of reference points is varied from $N_r = 25$ (for the MP2/cc-pVDZ method) to $N_r = 17$ [for CCSD(T)/(aug-)cc-pVDZ method]. To describe the instanton path properly by evenly distributing the reference points $\mathbf{q}_0(z_n)$, the first three values of z_n have to be chosen close to $z = -1$.

6.3 HOW TO USE THE THEORY

6.3.1 Evaluation of the Pre-Exponential Factor

Taking the malonaldehyde molecule as an example, we explain how to efficiently evaluate the pre-exponential factor B given by [see Equation (6.100)]

$$B = \sqrt{\frac{4g_\Sigma \det \mathbf{A}_m}{\pi g_0 \det \mathbf{A}_\Sigma}\frac{(\mathbf{p}^T\mathbf{g}\mathbf{p})_\Sigma}{\sqrt{(\mathbf{p}^T\mathbf{A}^{-1}\mathbf{p})_\Sigma}}}\exp(-2W_{1\Sigma}). \quad (6.117)$$

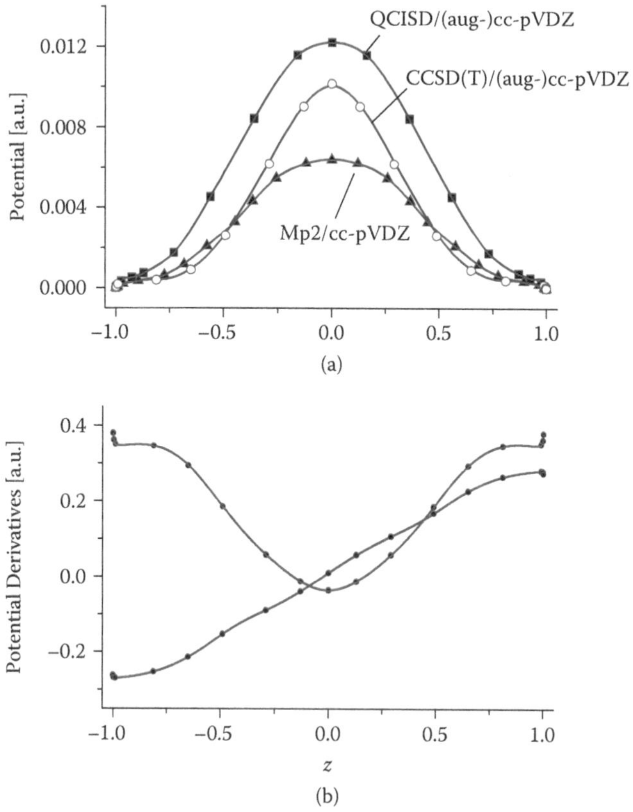

FIGURE 6.1 (a) One-dimensional interpolation of the potential $V[\mathbf{q}_0(z)]$ as a function of the parameter z along the instanton path $\mathbf{q}_0(z)$ for three *ab initio* methods in the case of malonaldehyde. The points represent the *ab initio* data. The lines are obtained by piecewise cubic interpolation. (b) One-dimensional interpolation of the elements of Hessian $\partial^2 V[\mathbf{q}(z)]/\partial q^i \partial q^j$ along the instanton path for MP2/cc-pVDZ *ab initio* method. Two examples are shown. (Taken from Reference [122] with permission.)

The first step is to define 21 internal coordinates of the nine-atomic malonaldehyde molecule. We assume the numbering of atoms as shown in Figure 6.2 and introduce the Cartesian coordinates x_{in} in the body-fixed frame of reference (BF) as

$$\mathbf{r}_n = x_{1n}\mathbf{e}_1 + x_{2n}\mathbf{e}_2 + x_{3n}\mathbf{e}_3, \tag{6.118}$$

where $n = (1, 2, \ldots, 9)$ numerates the atoms.

This BF frame is defined by the two conditions. The first one is to put the origin in the center of mass of the molecule,

$$\sum_{n=1}^{9} M_n x_{in} = 0, \quad i = 1, 2, 3, \tag{6.119}$$

where M_n is the mass of the nth atom. The second condtion fixes the orientation of the BF axis $\{\mathbf{e}_i\}$. We require that (1) the tunneling hydrogen atom H_9 lies in the $(\mathbf{e}_1, \mathbf{e}_2)$

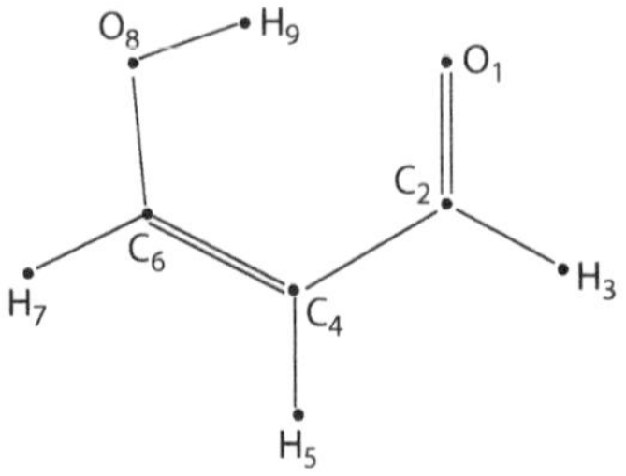

FIGURE 6.2 Malonaldehyde molecule. (Taken from Reference [123] with permission.)

plane and (2) $\mathbf{e}_1$ is directed along the line connecting the two oxygen atoms O_1 and O_8. In terms of x_{in} these two conditions read

$$x_{39} = 0, \quad x_{21} = x_{28}, \quad x_{31} = x_{38}. \tag{6.120}$$

Equations (6.119) and (6.120) give six constraints and define the BF system uniquely. The internal coordinates $q^k (k = 1, 2, \ldots, 21)$ are generally introduced in such a way that the functions $x_{in}(q^1, q^2, \ldots, q^{21})$ satisfy Equations (6.119) and (6.120) at any q^k. In our calculations we took 6 Cartesian coordinates—$x_{19}, x_{29}, x_{18}, x_{11}, x_{21}, x_{31}$—from the triangle O_8–H_9–O_1 and 15 Cartesian coordinates of 5 atoms C_2, C_4, C_6, H_3, H_7. We denote these 21 coordinates as $q^k (k = 1, 2, \ldots, 21)$. Their explicit numeration is immaterial. Equations (6.119) and (6.120) enable us to express the rest of the Cartesian coordinates (x_{39}, x_{28}, x_{38}, and, $\mathbf{r}_5$) in terms of $\{q^k\}$. With this definition, $x_{in}(q^1, q^2, \ldots, q^{21})$ are just linear combinations of $\{q^k\}$, which is convenient for constructing the metric tensor. Note also that the symmetry of the potential is easily established in terms of these internal coordinates. This naturally reduces the numerical effort to one-half.

Construction of the Hamiltonian for zero total angular momentum ($J = 0$) is now straightforward. We consider the kinetic metric

$$ds^2 = \sum_n M_n d\mathbf{r}_n^2 \tag{6.121}$$

and perform the transformation

$$\{d\mathbf{r}_n\} \longrightarrow \{dq^k, d\boldsymbol{\Omega}, d\mathbf{R}_{c.m.}\}, \tag{6.122}$$

where $d\boldsymbol{\Omega} = (d\omega_1, d\omega_2, d\omega_3)$ represent the projections of the infinitesimal rotation onto the BF axes and $\mathbf{R}_{c.m.}$ is the center-of-mass position. The latter separates from the equation of motion and does not need to be considered. Inserting $d\mathbf{r}_n = (\partial \mathbf{r}_n / \partial q^k) dq^k + [d\boldsymbol{\Omega} \times \mathbf{r}_n]$ into Equation (6.121), we obtain

$$\begin{aligned} ds^2 &= \sum M_n [dx_{in} dx_{in} + 2e_{ijk} x_{kn} dx_{in} d\omega^j + (\mathbf{r}_n^2 \delta_{ij} - x_{in} x_{jn}) d\omega^i d\omega^j] \\ &= \sum M_n \left(\frac{\partial x_{in}}{\partial q^k} \frac{\partial x_{in}}{\partial q^l} dq^k dq^l + 2e_{ijs} x_{sn} \frac{\partial x_{in}}{\partial q^l} dq^l d\omega^j \right. \\ &\quad \left. + (\mathbf{r}_n^2 \delta_{ij} - x_{in} x_{jn}) d\omega^i d\omega^j \right), \end{aligned} \tag{6.123}$$

where e_{ijk} is a fully antisymmetric tensor with respect to its three indices. Equation (6.123) explicitly determines the covariant metric tensor $G_{ij}(\mathbf{q})$ in terms of the 24 generalized coordinates $d\mathbf{Q} = \{d\mathbf{q}, d\boldsymbol{\Omega}\}$:

$$ds^2 = (d\mathbf{q}^T, d\boldsymbol{\Omega}^T) \begin{pmatrix} G_{qq} & G_{q\omega} \\ G_{\omega q} & G_{\omega\omega} \end{pmatrix} \begin{pmatrix} d\mathbf{q} \\ d\boldsymbol{\Omega} \end{pmatrix}. \tag{6.124}$$

With above choice of coordinates, the internal (G_{qq}), Coriolis ($G_{q\omega}$), and rotational ($G_{\omega\omega}$) parts of the metric are constant, linear, and quadratic functions of $\mathbf{q}$, respectively. The expression of the kinetic energy operator now reads

$$T = -\frac{\hbar^2}{2\sqrt{G}} \sum_{i,j=1}^{24} \frac{\partial}{\partial Q^i} \left(\sqrt{G} G^{ij} \frac{\partial}{\partial Q^j} \right), \tag{6.125}$$

where $\partial/\partial\mathbf{Q} = (\partial/\partial\mathbf{q}, \partial/\partial\boldsymbol{\Omega})$, G^{ij} are the elements of the inverse of the covariant metric tensor

$$\sum_{k=1}^{24} G_{ik} G^{kj} = \delta_{ij}, \tag{6.126}$$

and $G = |G_{ij}|$ is the determinant of the metric. Setting

$$\frac{\partial}{\partial\boldsymbol{\Omega}} \equiv \hat{\mathbf{J}} = 0 \tag{6.127}$$

in Equation (6.125), we obtain the kinetic-energy operator for zero angular momentum in the general form

$$\hat{H} = -\frac{\hbar^2}{2\sqrt{G}} \sum_{ij} \frac{\partial}{\partial q^i} \left(\sqrt{G} g^{ij} \frac{\partial}{\partial q^j} \right) + V(\mathbf{q}), \tag{6.128}$$

where g^{kl} is the 21×21 block of G^{ij}. Now Equation (6.97) can be solved.

6.3.2 Incorporation of High Level of *ab initio* Quantum Chemical Calculations

In order to build a nice bridge between theory and experiment, high accuracy of the potential data is definitely needed. Although accurate estimates of electronic states of large molecules are now possible with use of the high-level *ab initio* method, it is still a formidable task, as mentioned above, to obtain a *global* potential energy surface of such accuracy. Since calculations of the second derivatives of the potential at each point are also required, the whole quantum chemical calculations would become rather hopelessly formidable. It is desirable to reduce the total number of reference configurations. Since the tunneling splitting dynamics is determined essentially by the local property along the instanton path, and the global topology of the path does not strongly depend on the quality of the *ab initio* method, it is recommended to find the instanton path first by using a non-time-consuming low level of the *ab initio* method. In the first step of the calculations, the simple straight-line path $C^{in} = 0$

[see Equation (6.111)] is taken as an initial guess. This is actually unavoidable, since no information about the topography of the potential energy surface is known *a priori*. The number of iterations required for the convergence strongly depends on the initial guess for the instanton path, but the computational effort can be reduced by steadily improving the accuracy of the *ab initio* method and by using the converged instanton at a lower level of the *ab initio* method as the initial guess for the more accurate and time-consuming level of calculations (see Figure 6.1). By using this scheme, only a few extra steps are required to get convergence at the higher level.

In the case of a malonaldehyde molecule, as is described in more detail in Section 7.3, we use the MP2 method as the lowest level, and the quadratic configuration interaction method including singles and doubles substitutions (QCISD) and the coupled-cluster singles and doubles including a perturbational estimate of triple excitations [CCSD(T)] with the basis set of the Dunning's cc-pVDZ (Reference [124]) [CCSD(T)/(aug-)cc-pVDZ] as the intermediate levels. At the final step of the whole procedure, the CCSD(T) with the hybrid basis set of aug-cc-pVDZ method is used as the highest level. The number of iterations ($N_{\rm iter}$) required to obtain the converged instanton at higher *ab initio* levels was found to be only 4. It was also confirmed that $N_r = 17$ (number of reference points) is good enough to obtain smooth instanton path. Due to the symmetry of the molecule, the required number of *ab initio* calculations for $N_{\rm iter}$ iterations is $1 + 8N_{\rm iter} = 33$. The above-mentioned scheme together with the method of evaluation of the pre-exponential factor described in the previous subsection enables us to accurately estimate the tunneling splitting to be compared with the high-resolution spectroscopy experiment for virtually any polyatomic molecules.

6.4 CASE OF LOW VIBRATIONALLY EXCITED STATES

Tunneling splitting in the case of low vibrationally excited states is formulated in this section. The ordinary WKB theory explained in Section 3.1 cannot be extended easily to systems beyond two dimensions, because it is based on the complex-valued solutions of the Hamilton-Jacobi equation in the classically forbidden region. The instanton-type modified WKB theory can overcome this difficulty and can be formulated in canonically invariant form similarly to the case of ground-state tunneling splitting case in the multidimensional space. The formulation is made separately for longitudinal and transversal excitation with respect to the instanton trajectory. Furthermore, the multidimensional effects—i.e., nonmonotonic behavior of tunneling splitting against vibrational excitation explained in Chapter 4 based on the numerical solutions—can be interpreted by the behavior of effective frequencies along the instanton trajectory. In order to facilitate the understanding of the formulation, we start with the simplest one- and two-dimensional cases. The formulations in terms of the multidimensional Cartesian coordinate case and more general curved space case is provided after that.

6.4.1 ONE- AND TWO-DIMENSIONAL CASES

Let us summarize the main equations of Section 2.5.2, keeping the mass factor. We consider a symmetric one-dimensional potential with two equivalent potential minima at $x = \pm x_m$. If the potential barrier separating the two potential wells is high enough, the tunneling splitting can be calculated by the Herring formula,

$$\Delta = -\frac{2\hbar^2 \Psi_{sc}(0)\frac{d\Psi_{sc}}{dx}(0)}{mN}, \tag{6.129}$$

where $\Psi_{sc}(x)$ is the semiclassical wave function localized in the vicinity of one of the potential minima, m is the mass, and $N = \int \Psi_{sc}^2 dx$ is the normalization factor. We look for a real-valued semiclassical wave function in the form

$$\Psi_{sc} = \exp\left[-\frac{W_0}{\hbar} - W_1\right] \tag{6.130}$$

and treat the energy $E \propto \hbar$ as first-order term with respect to $\hbar$. Then, W_0 and W_1 satisfy the Hamilton-Jacobi equation and the transport (or continuity) equation, respectively, as

$$\frac{1}{2}\left(\frac{\partial W_0}{\partial x}\right)^2 - V(x) = 0 \tag{6.131}$$

and

$$\frac{1}{m}\frac{\partial W_0}{\partial x}\frac{\partial W_1}{\partial x} - \frac{1}{2m}\frac{\partial^2 W_0}{\partial x^2} + \frac{E}{\hbar} = 0. \tag{6.132}$$

These equations can be easily solved as

$$W_0(x) = \int_{-x_m}^{x} p_0(x)dx \tag{6.133}$$

and

$$W_1(x) = \int^{x}\left(\frac{1}{2}\frac{dp_0}{dx} - \frac{mE}{\hbar}\right)\frac{dx}{p_0}, \tag{6.134}$$

where $p_0(x) = \sqrt{2mV(x)}$ is the classical momentum along the instanton trajectory $x_0(\tau)$, $\tau \in [-\infty, \infty]$ that runs between the two potential minima. Note that the instanton trajectory is determined up to an arbitrary time shift. Without loss of generality, we always assume it to be symmetric in time, i.e., $x_0(\tau) = -x_0(\tau)$ and $x_0(0)$ is the position of the potential barrier top.

Since the principal exponential factor $W_0(x)$ does not depend on the energy E, the difference between the ground and excited states comes from the solution of the transport (continuity) equation. In the case of ground state, $E_{n=0} = \hbar\omega/2$, we note that the integrand in Equation (6.134) is well defined at $x = -x_m$, which can be taken as the low integration limit, and we have

$$W_1^{(n=0)} = \frac{1}{2}\int_{-x_m}^{x}\left(\frac{dp_0}{dx} - m\omega\right)\frac{dx}{p_0} = \frac{1}{2}\int_{-\infty}^{\tau}\left(\frac{1}{m}\frac{dp_0}{dx} - \omega\right)d\tau, \tag{6.135}$$

where n represents the vibrational quantum number.

For the first excited state we put $E_{n=1} = E_{n=0} + \hbar\omega$ and look for the solution of the transport (continuity) equation in the form

$$W_1^{(n=1)} = W_1^{(n=0)} + w, \tag{6.136}$$

where the additional term w satisfies the equation

$$\frac{p_0(x)}{m}\frac{dw(x)}{dx} + \omega = 0. \tag{6.137}$$

There is a simple singularity at $x = -x_m$ in this equation that leads to the logarithmic divergence of $w(x)$ and we have

$$w(x) = -\int^x \frac{m\omega}{p_0(x)}dx = -\int^\tau \omega d\tau = \int_{-\infty}^{\tau}\left(\frac{1}{m}\frac{dp_0}{dx} - \omega\right)d\tau - \ln p_0(x). \tag{6.138}$$

As a result, for $x \to -x_m$, $\exp[-w(x)] \simeq p_0(x)$ and the semiclassical wave function

$$\Psi_{sc} = p_0(x)\exp[-m\omega(x+x_m)^2/2\hbar] \simeq m\omega(x+x_m)\exp[-m\omega(x+x_m)^2/2\hbar] \tag{6.139}$$

coincides with the excited state. Equations (6.133), (6.135), and (6.138) give the semiclassical wave function for $n = 1$. The tunneling splitting $\Delta_{n=1}$ is obtained from the Herring formula. Taking into account the difference between the normalization integrals $N_{n=1} = (\hbar m\omega/2)N_{n=0}$, we obtain

$$\Delta_{n=1} = 2\Delta_{n=0}\frac{p_0^2(0)}{\hbar m\omega}\exp\left[2\int_{-\infty}^{0} d\tau\left(\omega - \frac{1}{m}\frac{dp_0}{dx}\right)\right]. \tag{6.140}$$

Note that for a double well potential, dp_0/dx is a monotonically decreasing function of time and the barrier height $V(0)$ always exceeds the excitation energy. Thus,

$$\Delta_{n=1} > \Delta_{n=0} \tag{6.141}$$

always holds. See also the discussion after Equation (6.160). As is seen later, Equation (6.140) can be used for the case of longitudinal excitation in a multidimensional system.

Let us next consider the simple two-dimensional case [125]. Here we use the local coordinates (s, ξ) where $s \in [-s_m, s_m]$ runs along the instanton trajectory $\mathbf{q}_0(s)$ and ξ is the coordinate perpendicular to the instanton. The solution of the Hamilton-Jacobi equation in the vicinity of the instanton trajectory $\xi = 0$ reads [see Equation (6.67)]

$$W_0(s,\xi) = \int_{-s_m}^{s} p_0(s')ds' + \frac{m\theta(s)}{2}\xi^2, \tag{6.142}$$

where $p_0(s) = \sqrt{2mV(\mathbf{q}_0(s))}$ is the classical momentum along the instanton trajectory and $\theta(s)$ plays a role of the effective frequency in the direction of the transversal local coordinate ξ. As before, the difference between the ground and excited states comes from the energy factor in the transport (continuity) equation. Keeping only the relevant terms $o(\xi)$, it can be written as

$$\frac{p_0(s)}{m}\frac{\partial W_1}{\partial s} + \theta(s)\xi\frac{\partial W_1}{\partial \xi} - \frac{1}{2m}\frac{\partial p_0}{\partial s} - \frac{1}{2}\theta(s) + \frac{E_0 + \Delta E}{\hbar} = 0, \tag{6.143}$$

where $E_0 = \hbar(\omega_\parallel + \omega_\perp)/2$ is the ground-state energy with $\omega_\parallel$ and $\omega_\perp$ being normal frequencies for the longitudinal and transversal modes, respectively, and ΔE is the excitation energy. The solution of Equation (6.143) is given again in the form $W_1^{(n=1)} = W_1^{(n=0)} + w$ [see Equation (6.136)] where

$$W_1^{(n=0)} = \frac{1}{2}\int_{-s_m}^{s}\left(\frac{1}{m}\frac{dp_0}{ds'} + \theta(s') - \omega_\| - \omega_\perp\right)\frac{m}{p_0(s')}ds' \tag{6.144}$$

and the additional term $w(s, \xi)$ satisfies the equation

$$\frac{p_0(s)}{m}\frac{\partial w}{\partial s} + \theta(s)\xi\frac{\partial w}{\partial \xi} + \frac{\Delta E}{\hbar} = 0. \tag{6.145}$$

In the case of longitudinal excitation ($n_\| = 1, n_\perp = 0$), we take $\Delta E = \hbar\omega_\|$, $w = w(s)$, and obtain the solution totally identical to the previous one-dimensional case, Equation (6.138).

For the transversal excitation, we take $\Delta E = \hbar\omega_\perp$ and obtain

$$w(s, \xi) = \int_{-s_m}^{s}[\theta(s') - \omega_\perp]\frac{m}{p_0(s')}ds' - \ln\xi. \tag{6.146}$$

In this case the semiclassical wave function Equation (6.130) in the vicinity of the potential minimum $(-s_m, 0)$ becomes

$$\Psi_{sc} = \xi\exp\left(-\frac{m\omega_\|(s + s_m)^2}{2\hbar} - \frac{m\omega_\perp\xi^2}{2\hbar}\right), \tag{6.147}$$

which coincides with the excited-state wave function for ($n_\| = 0, n_\perp = 1$) of the two-dimensional harmonic oscillator. Taking into account the normalization factor $N_{n_\perp=1} = \hbar N_0/(2m\omega_\perp)$, we obtain

$$\Delta_{n_\perp=1} = \Delta_{n=0}\frac{\omega_\perp}{\theta(0)}\exp\left[2\int_{-\infty}^{0}d\tau(\omega_\perp - \theta)\right]. \tag{6.148}$$

As mentioned before, the longitudinal excitation always promotes tunneling, $\Delta_{n_\|=1} > \Delta_{n=0}$. The effect of transversal excitation, however, depends on the behavior of the effective frequency $\theta(\tau)$. For instance, for monotonically growing (decreasing) $\theta(\tau)$ the excitation of the transversal mode suppresses (promotes) the tunneling splitting. We see below that the effective frequency is generally determined as a solution of auxiliary differential equation that describes the nodal structure of the semiclassical wave function. This corresponds to the discussion of Takada and Nakamura [30,31] (see Chapter 4).

6.4.2 Multidimensional Case in Terms of Cartesian Coordinates

Here we consider an N-dimensional problem with the Hamiltonian

$$H = \frac{\mathbf{p}^2}{2} + V(\mathbf{x}), \tag{6.149}$$

where $V(x)$ is a symmetric double well potential. As usual, we introduce the $(N-1)$-dimensional dividing surface Σ, which has a simple meaning of location of the potential barrier separating two equivalent potential wells and can be defined by the algebraic equation $f(\mathbf{x}) = 0$. Hereafter we put the Planck constant $\hbar = 1$. The multidimensional Herring formula assumes the form

$$\Delta = -2\frac{\int d\mathbf{x}\delta(f)\Psi_{sc}\sum_i \frac{\partial \Psi_{sc}}{\partial x_i}\frac{\partial f}{\partial x_i}}{\int d\mathbf{x}\Psi_{sc}^2}, \tag{6.150}$$

where Ψ_{sc} is the semiclassical wave function with the asymptotic form of Equation (6.130). The corresponding Hamilton-Jacobi equation and the transport (continuity) equation read

$$\frac{1}{2}\sum_i \left(\frac{\partial W_0}{\partial x_i}\right)^2 - V(\mathbf{x}) = 0 \tag{6.151}$$

and

$$\sum_i \left[\frac{\partial W_0}{\partial x_i}\frac{\partial W_1}{\partial x_i} - \frac{1}{2}\frac{\partial^2 W_0}{\partial x_i^2}\right] + E = 0. \tag{6.152}$$

The solution $W_0(\mathbf{x})$ is given by the action integral, as usual, along the characteristics, i.e., classical trajectories on the upside-down potential with zero energy starting from the potential minimum $\mathbf{x}_m$. The instanton trajectory $\mathbf{x}_0(\tau)$ is one of the characteristics that connects $\mathbf{x}_m$ to the other potential minimum. The instanton trajectory crosses the dividing surface at $\tau = 0$ at $\mathbf{x}_\Sigma = \mathbf{x}_0(\tau = 0)$. The problem is thus reduced to solving Equations (6.151) and (6.152) in the vicinity of $\mathbf{x}_0(\tau)$. The Hamilton-Jacobi equation, Equation (6.151), does not depend on energy and it was solved already in Section 6.1.2. For the present purpose, we need an explicit expression of $W_0(\mathbf{x})$ in the vicinity of the instanton trajectory.

For this purpose, let us introduce a function $\tau(\mathbf{x})$ according to the orthogonality relation

$$\sum_i [x_i - x_{0i}(\tau)]\dot{x}_{oi}(\tau) = 0. \tag{6.153}$$

Then it can be shown that $W_0(\mathbf{x})$ is given by

$$W_0(\mathbf{x}) = \sum_i \int_{-\infty}^{\tau} p_{0i}(\tau')\dot{x}_{0i}(\tau')d\tau'$$
$$+\frac{1}{2}\sum_{ij} A_{ij}(\tau)[x_i - x_{0i}(\tau)][x_j - x_{0j}(\tau)] + o[(\Delta\mathbf{x})^2], \tag{6.154}$$

where the time τ should be understood as a function of $\mathbf{x}$ and the symmetric matrix $\mathbf{A}(\tau) \equiv \partial^2 W_0/\partial \mathbf{x}^2|_{\mathbf{x}_0(\tau)}$ satisfies the equation [see Equation (6.75)]

$$\frac{d}{d\tau}\mathbf{A} = V_{\mathbf{xx}}[\mathbf{x}_0(\tau)] - \mathbf{A}^2(\tau). \tag{6.155}$$

Indeed, by differentiating Equation (6.153) we find (see Appendix F)

$$\frac{\partial \tau}{\partial \mathbf{x}} = \frac{\mathbf{v}_0}{\mathbf{v}_0^2 - (\mathbf{a}\Delta\mathbf{x})} = \frac{\mathbf{v}_0}{\mathbf{v}_0 \cdot \mathbf{v}_0}\left(1 + \frac{\mathbf{a}\Delta\mathbf{x}}{\mathbf{v}_0^2} + \frac{(\mathbf{a}\Delta\mathbf{x})^2}{(\mathbf{v}_0^2)^2} + o\left((\Delta\mathbf{x})^2\right)\right), \tag{6.156}$$

where $\mathbf{v}_0$ is the velocity of the instanton, $\mathbf{a} \equiv \dot{\mathbf{v}}_0 = \nabla V$ and $\Delta\mathbf{x} = \Delta\mathbf{x} - \mathbf{x}_0(\tau)$. Note also the useful relation (see Appendix F)

$$a_i(\tau) = \sum_j A_{ij} v_{oj}(\tau). \tag{6.157}$$

Using Equations (6.155)–(6.157), one can easily check that Equation (6.154) satisfies the Hamilton-Jacobi equation up to the terms $o[(\Delta\mathbf{x})^2]$. At $\tau = -\infty$, the left side of Equation (6.155) vanishes and $\mathbf{A}(-\infty) \equiv \mathbf{A}_m$ becomes a matrix of the normal frequencies at the potential minimum $\mathbf{x}_m$.

Now we can proceed with the transport equation. In the case of ground state, $W_1^{(n=0)}$ can be solved easily by integrating Equation (6.152) along the instanton trajectory as

$$W_1^{(n=0)} = \frac{1}{2}\int_{-\infty}^{\tau} \left[\mathrm{Tr}[\mathbf{A}(\tau') - \mathbf{A}_m]\right] d\tau'. \tag{6.158}$$

We saw in the previous subsection that $W_1^{(n=1)}$ for the first excited state differs from $W_1^{(n=0)}$ by the logarithmically divergent term. In the N-dimensional case, this correction term w [see Equation (6.136)] satisfies the equation

$$\sum_i \frac{\partial W_0}{\partial x_i}\frac{\partial w}{\partial x_i} + \Delta E = 0. \tag{6.159}$$

First, we note that the instanton trajectory can be parametrized by some longitudinal "tunneling" coordinate s that coincides with one of the normal coordinates at the potential minimum. This longitudinal excitation is equivalent to the one-dimensional problem and the corresponding tunneling splitting is given by Equation (6.148) in the previous section. We can further rewrite Equation (6.148) in the invariant form by using the time τ along the instanton as the integration variable. This gives the following expression, which can be directly implemented for arbitrary multidimensional system:

$$\Delta_{n=1} = \Delta_{n=0}\frac{4V(0)}{\omega}\exp\left[2\int_{-\infty}^{0} d\tau\left(\omega - \frac{1}{2V}\frac{dV}{d\tau}\right)\right], \tag{6.160}$$

where $dp_0/dx = (m/p_0)dp_0/d\tau = (m/2V)dV/d\tau$ is used. Since $V(0) > \hbar\omega/4$ and the exponent is positive, Equation (6.141) holds, as mentioned before.

In the case of transversal excitation, we look for $w(\mathbf{x})$ in the form

$$w = \int [\theta(\tau') - \Delta E]d\tau' - \ln(\mathbf{U}^T(\tau)\Delta\mathbf{x}), \tag{6.161}$$

where, again, τ is understood as a function of $\mathbf{x}$ [see Equation (6.156)]. Inserting $\partial W_0/\partial x_i$ and $\partial w/\partial x_i$ obtained with use of Equation (6.156) into Equation (6.159), multiplying by $\mathbf{U}\Delta\mathbf{x}$ and omitting the terms of order $o(\Delta\mathbf{x})$, we have (see Appendix F)

$$\theta(\tau)(\mathbf{U}^T\Delta\mathbf{x}) - (\mathbf{U}^T\mathbf{A}\Delta\mathbf{x}) - (\dot{\mathbf{U}}^T\Delta\mathbf{x}) + \frac{2(\mathbf{v}_0^T\mathbf{A}\Delta\mathbf{x})}{\mathbf{v}_0^2}(\mathbf{U}^T\mathbf{v}_0) = 0. \tag{6.162}$$

The initial condition $\mathbf{U}(-\infty) \equiv \mathbf{U}_m$ for Equation (6.162) must be chosen according to the desired behavior of the wave function at the potential minimum. Let γ specify the excited normal mode $\Delta E = \hbar\omega_\gamma$. Then, at $\tau \to -\infty$, $\dot{\mathbf{U}} = \mathbf{v}_0 = 0$, the last two terms in Equation (6.162) vanish, and we find that $\theta(-\infty)$ coincides with the normal frequency ω_γ and $\mathbf{U}_m$ is the corresponding eigenvector of $\mathbf{A}_m$. One of the normal modes corresponds to the longitudinal excitation and the corresponding tunneling splitting is given by Equation (6.160). We see below that for the $N-1$ transversal modes $\mathbf{U}(\tau)$ satisfies the orthogonality condition,

$$(\mathbf{U}^T\mathbf{v}_0) = 0. \tag{6.163}$$

Hence, the last term in Equation (6.162) vanishes and we come to a simple equation

$$\dot{U}_i = \theta(\tau)U_i - \sum_j A_{ij}U_j. \tag{6.164}$$

Equation (6.163) does not determine $(\mathbf{U}, \theta)$ uniquely yet. Indeed, one can see that the solution is defined only up to the transformation $\mathbf{U} \to g\mathbf{U}, \theta \to \theta + (1/g)(dg/d\tau)$, where $g(\tau)$ is an arbitrary function. This can also be seen directly from the definition, Equation (6.161). To specify the solution of Equation (6.164), we impose the normalization condition

$$(\mathbf{U}^T\mathbf{U}) = 1. \tag{6.165}$$

Then, multiplying Equation (6.164) by $\mathbf{U}^T$ from the left, we find

$$\theta = (\mathbf{U}^T\mathbf{A}\mathbf{U}) \tag{6.166}$$

and finally obtain

$$\dot{U}_i = (\mathbf{U}^T\mathbf{A}\mathbf{U})U_i - \sum_j A_{ij}U_j. \tag{6.167}$$

Equations (6.161), (6.166), and (6.167) completely determine the term w in the semiclassical wave function for the first excited state of the transversal mode γ. As shown in Appendix F, one can easily verify that the orthogonality and normalization conditions, Equations (6.163) and (6.165), are consistent with Equation (6.167).

Now, we can proceed with the calculation of the tunneling splitting by the Herring formula, Equation (6.150). Noting that the normalization factor in the case of γ mode is equal to $N_{n_\gamma=1} = N_{n=0}/2\omega_\gamma$, expanding the δ function in the numerator of the Herring formula, and using the fact that $\nabla F \propto \mathbf{v}_0$ at the stationary point $\mathbf{x}_\Sigma$, we can obtain the following expressions for the ground- and excited-state tunneling splitting as

$$\Delta_{n=0} = \frac{2\mathbf{v}_0^2}{N_{n=0}} \exp(-S_0 - S_1) \int d\mathbf{x}\delta\left(\sum_i v_{0i}x_i\right) \exp(-\mathbf{x}^T\mathbf{A}\mathbf{x}) \tag{6.168}$$

and

$$\Delta_{n_\gamma=1} = \frac{2\mathbf{v}_0^2}{N_{n_\gamma=1}} \exp(-S_0 - S_1 - \Delta S_1)$$
$$\times \sum_{lm} \int d\mathbf{x}\delta\left(\sum_i v_{0i}x_i\right) U_l U_m x_l x_m \exp(-\mathbf{x}^T\mathbf{A}\mathbf{x}), \tag{6.169}$$

where S_0 is the action along the instanton, $S_1 = 2W_1^{(n=0)}(0)$ and ΔS_1 is the correction due to the first term of Equation (6.161),

$$\Delta S_1 = 2\int_{-\infty}^{0} [\theta(\tau) - \omega_\gamma]d\tau. \tag{6.170}$$

In the above equations it is understood that $\mathbf{v}_0$, $\mathbf{A}$, and $\mathbf{U}$ are taken at $\tau = 0$, i.e., at the crossing point $\mathbf{x}_\Sigma$. To evaluate the integrals, we introduce the generating function

$$F(\xi) \equiv \int d^N\mathbf{x}\delta(\mathbf{v}_0^T\mathbf{x})\exp(-\mathbf{x}^T\mathbf{A}\mathbf{x} + i\xi^T\mathbf{x}) \tag{6.171}$$

and rewrite Equations (6.168) and (6.169) in the form

$$\Delta_{n=0} = \frac{2\mathbf{v}_0^2}{N_{n=0}}\exp(-S_0 - S_1)F(0) \tag{6.172}$$

and

$$\Delta_{n_\gamma=1} = -\frac{4\omega_\gamma\mathbf{v}_0^2}{N_{n=0}}\exp(-S_0 - S_1 - \Delta S_1)\sum_{lm} U_l U_m \frac{\partial^2 F(\xi)}{\partial\xi_l\partial\xi_m}|_{\xi=0}. \tag{6.173}$$

$F(\xi)$ can be calculated by making the Fourier transformation of the δ function,

$$F(\xi) = \sqrt{\frac{\pi^{N-1}}{\det\mathbf{A}(\mathbf{p}_0^T\mathbf{A}^{-1}\mathbf{p}_0)}}\exp\left[-\frac{1}{4}\xi^T\mathbf{A}^{-1}\xi + \frac{(\xi^T\mathbf{A}^{-1}\mathbf{p}_0)^2}{4(\mathbf{p}_0^T\mathbf{A}^{-1}\mathbf{p}_0)}\right], \tag{6.174}$$

and we finally obtain

$$\Delta_{n_\gamma=1} = \Delta_{n=0}\omega_\gamma\left(\mathbf{U}^T\left[\mathbf{A}^{-1} + \frac{(\mathbf{A}^{-1}\mathbf{p}_0)\otimes(\mathbf{p}_0^T\mathbf{A}^{-1})}{(\mathbf{p}_0^T\mathbf{A}^{-1}\mathbf{p}_0)}\right]\mathbf{U}\right)_\Sigma\exp(-\Delta S_1). \tag{6.175}$$

This expression gives the tunneling splitting for a single excitation of the transversal mode γ. This result is invariant under any linear coordinate transformation. We generalize the theory to the case of arbitrary curvilinear coordinates in the next subsection. The case of multiple excitation is also discussed briefly.

6.4.3 Case of General Multidimensional Curved Space

We now consider the general Hamiltonian, which is the same as Equation (6.82) in Subsection 6.1.3,

$$H = -\frac{1}{2\sqrt{g}}\frac{\partial}{\partial q^i}\left(\sqrt{g}g^{ij}\frac{\partial}{\partial q^j}\right) + V(\mathbf{q}), \tag{6.176}$$

where g^{ij} is the metric tensor and $g = ||g_{ij}||$. There are some duplications of equations used in the previous sections, but we repeat them for the self-sufficiency of this subsection. The potential $V(\mathbf{q})$ is assumed to have two equivalent minima located at $\mathbf{q}_m$ and $\tilde{\mathbf{q}}_m$. Since the formulation is similar to that in the previous subsection, only the outline is provided here. The Hamilton-Jacobi equation now reads [see Equation (6.86)]

$$H\left(\frac{\partial W_0}{\partial \mathbf{q}}, \mathbf{q}\right) = 0, \tag{6.177}$$

where

$$H(\mathbf{p}, \mathbf{q}) = \frac{1}{2}\sum_{ij} g^{ij} p_i p_j - V(\mathbf{q}) \tag{6.178}$$

is the classical Hamiltonian with upside-down potential [see Equation (6.88)]. We define the function $\tau(\mathbf{q})$ according to the relation [see Equation (6.153)]

$$\sum_i [q^i - q_0^i(\tau)] p_{0i}(\tau) = 0, \tag{6.179}$$

where $\mathbf{p}_0(\tau)$ is the momentum of the instanton trajectory. Then it can be shown that in the vicinity of the instanton trajectory the solution of the Hamilton-Jacobi equation can be expressed in the form [see Equation (6.154)]

$$W_0 = \int_{-\infty}^{\tau} \sum_i p_{0i}(\tau')\dot{q}_0^i(\tau')d\tau' + \frac{1}{2}\sum_{ij} \tilde{A}_{ij}(\tau)[q^i - q_0^i(\tau)]\left[q^j - q_0^j(\tau) + o\left((\Delta\mathbf{q})^2\right)\right], \tag{6.180}$$

where τ is understood as a function of $\mathbf{q}$ and the symmetric matrix $\tilde{\mathbf{A}}(\tau)$ satisfies the equation [see Equation (6.97)]

$$\dot{\tilde{\mathbf{A}}} = -\tilde{\mathbf{A}}H_{\mathbf{pp}}\tilde{\mathbf{A}} - H_{\mathbf{qq}} - \tilde{\mathbf{A}}H_{\mathbf{pq}} - H_{\mathbf{qp}}\tilde{\mathbf{A}}. \tag{6.181}$$

All the quantities in this equation are taken along the instanton trajectory and $\tilde{\mathbf{A}}(\tau)$ has the same meaning of the Hessian matrix of the classical action on the instanton trajectory. Note that Equation (6.155) in the previous subsection is a particular case of this general expression for the unit metric tensor.

The transport (continuity) equation in the general case reads [see Equation (6.87)]

$$\sum_{ij}\left(g^{ij}\frac{\partial W_0}{\partial q^i}\frac{\partial W_1}{\partial q^j} - \frac{1}{2}g^{ij}\frac{\partial^2 W_0}{\partial q^i \partial q^j}\right) + E - \sum_i \frac{1}{2}\Lambda^i\frac{\partial W_0}{\partial q^i} = 0, \tag{6.182}$$

where Λ is defined by the same equation as Equation (6.89)

$$\Lambda^i = \frac{1}{\sqrt{g}}\sum_j \frac{\partial}{\partial q^j}(\sqrt{g}g^{ij}). \tag{6.183}$$

For the ground state we have

$$E_{n=0} = \frac{1}{2}\sum_{\gamma}\omega_{\gamma} = \frac{1}{2}\mathrm{Tr}\tilde{\mathbf{A}}_m \tag{6.184}$$

and Equation (6.182) is solved by integrating along the instanton trajectory as [see Equation (6.105)]

$$W_1^{(n=0)} = \int_{-\infty}^{\tau}\left(\frac{1}{2}\mathrm{Tr}[\tilde{\mathbf{A}}(\tau') - \tilde{\mathbf{A}}_m] + \frac{1}{2}\mathbf{p}_0^T(\tau')\mathbf{\Lambda}(\tau')\right)d\tau', \tag{6.185}$$

where

$$\text{Tr}\tilde{\mathbf{A}} \equiv \sum_i \tilde{\mathbf{A}}_i^i = \sum_{ij} g^{ij}\tilde{\mathbf{A}}_{ij}. \tag{6.186}$$

This result can be rewritten in a more compact form as

$$W_1^{(n=0)} = \frac{1}{2}\int_{-\infty}^{\tau} \text{Tr}[\mathbf{A}(\tau') - \mathbf{A}_m]d\tau' \tag{6.187}$$

with $\mathbf{A}$ defined by [see Equations (6.101) and (6.103)]

$$A_{ij} = \tilde{A}_{ij} - \sum_k p_{0k}\Gamma_{ij}^k, \tag{6.188}$$

where Γ is given by [see Equation (6.104)]

$$\Gamma_{ij}^k = \frac{1}{2}\sum_s g^{ks}\left(\frac{\partial g_{is}}{\partial q^i} + \frac{\partial g_{js}}{\partial q^j} - \frac{\partial g_{ij}}{\partial q^s}\right), \tag{6.189}$$

which are the Christoffel symbols. Note that unlike $\tilde{\mathbf{A}}$ the "full" matrix $\mathbf{A}$ transforms as a tensor under arbitrary nonlinear coordinate transformation. We showed in Subsection 6.1.3 that the tunneling splitting of the ground state is expressed in terms of $\mathbf{A}$ rather than $\tilde{\mathbf{A}}$, which makes the final results independent of the coordinate representation. For the low excited states, the tunneling splitting can also be expressed in the similar invariant form.

The solution of the transport (continuity) equation differs from Equation (6.185) by the singular correction term $w(\mathbf{q})$, which is to be found from the equation

$$\sum_{ij} g^{ij}\frac{\partial W_0}{\partial q^i}\frac{\partial w}{\partial q^j} + \Delta E = 0. \tag{6.190}$$

Taking $w(\mathbf{q})$ in the form of Equation (6.161) we arrive at the equation

$$\dot{U}_k = \theta(\tau)U_k - \sum_{ij}[g^{ij}\tilde{A}_{ik} + \frac{\partial g^{ij}}{\partial q^k}p_{0i}]U_j. \tag{6.191}$$

We can specify the solution by imposing the normalization condition [see Equation (6.165)]

$$\sum_{ij} g^{ij}U_iU_j = 1 \tag{6.192}$$

and then we obtain [see Equation (6.166)]

$$\theta(\tau) = \sum_{ij} U^i A_{ij} U^j. \tag{6.193}$$

The covariant (lower index) and contravariant (upper index) vectors in the above equations are related in the usual way through the metric tensor as

$$U^i = \sum_k g^{ik}U_k. \tag{6.194}$$

The vector **U** characterizes the direction along which the semiclassical wave function has a node. At the potential minimum ($\tau = -\infty$) **U** coincides with one of the normal modes while $\theta(-\infty)$ is the corresponding normal mode frequency (excitation energy). Equation (6.191) has $N - 1$ independent solutions corresponding to $N - 1$ possible types of the transversal excitation. The fact that **U** indeed has the vector transformation properties can be deduced from Equation (6.191). Note also that similarly to Equation (6.164), this equation is pertinent only to the transversal excitation for which Equation (6.163) holds.

Equations (6.161), (6.191), and (6.193) completely determine the semiclassical wave function and the tunneling splitting can now be calculated by the Herring formula [see Equation (6.84)]

$$\Delta = \frac{2\int \sqrt{g} d\mathbf{q}\delta(f)\Psi \sum_{ij} g^{ij} \frac{\partial \Psi}{\partial q^i}\frac{\partial f}{\partial q^j}}{\int \sqrt{g} d\mathbf{q}\Psi^2}. \tag{6.195}$$

The normalization factor in the denominator is the same as before. When expanding the argument of the δ function, one generally has to take into account the curvature of the dividing surface so that the integral in the numerator is proportional to

$$\int d\mathbf{q}\delta\left(\sum_i \frac{\partial f}{\partial q^i} q^i + \frac{1}{2}\sum_{ij} \frac{\partial^2 f}{\partial q^i \partial q^j}\right)(\mathbf{U}^T\mathbf{q})^2 \exp(-2\mathbf{p}^T\mathbf{q} - \mathbf{q}^T\tilde{\mathbf{A}}\mathbf{q}). \tag{6.196}$$

The properties of the function $f(\mathbf{q})$ were discussed in Subsection 6.1.3. It was shown that at the stationary point $\mathbf{q}_\Sigma$ it satisfies the relations

$$\frac{\partial f}{\partial q^i} \propto p_i \text{ and } \frac{\partial^2 f}{\partial q^i \partial q^j} = \sum_k \frac{\partial f}{\partial q^k}\Gamma^k_{ij}. \tag{6.197}$$

Taking this into account and evaluating the integral the same way as in the previous subsection, we obtain the tunneling splitting in the form of Equation (6.175) with the full matrix **A** given by Equation (6.188). Note that this final formula is invariant under any coordinate transformation as it should be.

Equations (6.160) and (6.175) give the tunneling splitting for excited states $[n_\gamma = 1, n_{\gamma'} = 0](\gamma \neq \gamma')$ with a single mode excitation γ. Generalization to the case of M-mode excitation $(M < N)[n_{\gamma_k} = 1(k = 1, 2, \ldots, M), n_{\gamma'} = 0(\gamma' \neq \gamma)]$ is straightforward. Indeed, since the correction factor $w(\mathbf{q})$ satisfies the linear equation (6.190) and in the harmonic approximation

$$\Delta E = \sum_{k=1}^{M} \omega_{\gamma_k}, \tag{6.198}$$

we have

$$\exp[-w(\mathbf{q})] = \Pi_k \exp[-w_{\gamma_k}(\mathbf{q})], \tag{6.199}$$

where for each k, $w_{\gamma_k}(\mathbf{q})$ is given by either Equation (6.138) or Equation (6.161), this completely determines the wave function, and the tunneling splitting can be calculated from the Herring formula, Equation (6.195). For example, for M excited transversal modes $\gamma_1, \gamma_2, \ldots, \gamma_M$, we have generally

$$\frac{\Delta(\gamma_1, \gamma_2, \ldots, \gamma_M)}{\Delta_{(n=0)}} = \frac{\Pi_{k=1}^{M}(2\omega_{\gamma_k})}{F(0)} \exp\left(-\sum_{k=1}^{M} \Delta S_1^{(\gamma_k)}\right)$$
$$\times \sum U_{l_1}^{(\gamma_1)} U_{m_1}^{(\gamma_1)} U_{l_2}^{(\gamma_2)} U_{m_2}^{(\gamma_2)} \cdots U_{l_M}^{(\gamma_M)} U_{m_M}^{(\gamma_M)}$$
$$\times \frac{\partial^{2M} F(\xi)}{\partial \xi_{l_1} \partial \xi_{m_1} \partial \xi_{l_2} \partial \xi_{m_2} \cdots \partial \xi_{l_M} \partial \xi_{m_M}} |_{\xi=0}, \qquad (6.200)$$

where $F(\xi)$ is defined by Equation (6.174).

The invariant form of the obtained results is crucial for practical implementation of the theory, since it enables us to use any convenient internal coordinates to describe the molecular motion. One can see that the only difference from the ground-state problem is due to the matrix equation (6.191) for the evolution of the excitation along the instanton. Note that this equation involves only the matrix $\mathbf{A}(\tau)$, which describes the Gaussian shape of the semiclassical wave function of the ground state in the vicinity of the instanton trajectory $\mathbf{q}_0(\tau)$. For a given potential energy surface both $\mathbf{q}(\tau)$ and $\mathbf{A}(\tau)$ can be calculated very easily [43] and Equation (6.191) also presents no serious numerical difficulties. In general, the global analytical potential function is not available and time-consuming *ab initio* quantum chemical calculations are required. The method described in Section 6.3 can facilitate the *ab initio* part of the simulation and can be efficiently used for realistic systems of spectroscopic interest. The only auxiliary Equation (6.191) that is needed in the case of excited states does not require any additional *ab initio* data.

7 Numerical Applications to Polyatomic Molecules

7.1 N-DIMENSIONAL SEPARABLE POTENTIAL MODEL

Before dealing with real polyatomic molecules, let us first check the theory presented in Chapter 6 and explain the numerical procedure by taking a simple exactly solvable N-dimensional model. We consider a particle of mass m in the N-dimensional separable potential

$$V(\mathbf{x}) = V_0(x_1^2 - x_0^2)^2 + \sum_{i=2}^{N} \frac{m}{2}\omega_i^2 x_i^2. \tag{7.1}$$

In this case the instanton trajectory is just a straight line connecting the points $x_1 = \pm x_0$ with $x_i = 0 (i = 2, \ldots, N)$, and the tunneling splitting Δ_0 is given by the one-dimensional answer for the quartic potential $V_0(x_1^2 - x_0^2)^2$ (see Section 2.5). To test the present theory we perform the following nonlinear coordinate transformation $\mathbf{x} \rightarrow \mathbf{q}$:

$$q^k = T_{k1}x_1 + \sum_{i=2}^{N} T_{ki}(x_i + 1 + x_1^2)^2, \quad (k = 1, 2, \ldots, N), \tag{7.2}$$

where T_{ij} is the (i, j) element of the following $N \times N$ matrix T.

$$T = T^{1n}T^{1(n-1)} \cdots T^{12}, \tag{7.3}$$

where T^{1j} is also an $N \times N$ orthogonal matrix and represents the rotational matrix in the (x_1, x_j) plane. In other words, the diagonal elements of T^{1j} are all unity except for the 1st and the jth elements, which are $\cos(\alpha)$, and the off-diagonal elements are all zero except for the $(T^{1j})_{1j} = \sin(\alpha)$ and $(T^{1j})_{j1} = -\sin(\alpha)$, where α is the rotation angle. In the new coordinates $\mathbf{q}$ the Hamiltonian operator has the form of Equation (6.82) and the corresponding tunneling splitting Equation (6.100) is invariant under any coordinate transformation. The numerical computations in terms of the new coordinates allow checking the numerical procedure by comparison with the analytical answer. It should be noted that all the factors in Equation (6.100),

$$\frac{g}{\det \mathbf{A}}, \quad (\mathbf{p}^T\mathbf{A}^{-1}\mathbf{p}), \quad W_{0\Sigma}, \quad W_{1\Sigma}, \tag{7.4}$$

are invariant separately so that the numerical comparison can be made separately for each term.

The parameters used are $\alpha = 0.1$, $\omega_i = 1000\ \mathrm{cm}^{-1}$, $V_0 = 10$ Kcal/mol, and m is the proton mass. The calculations were carried out for $N = 4$ and $N = 20$. The accuracy was found to be the same for both cases and the results only for $N = 20$ are given here [43]. Figure 7.1 shows the iterative calculations of the instanton

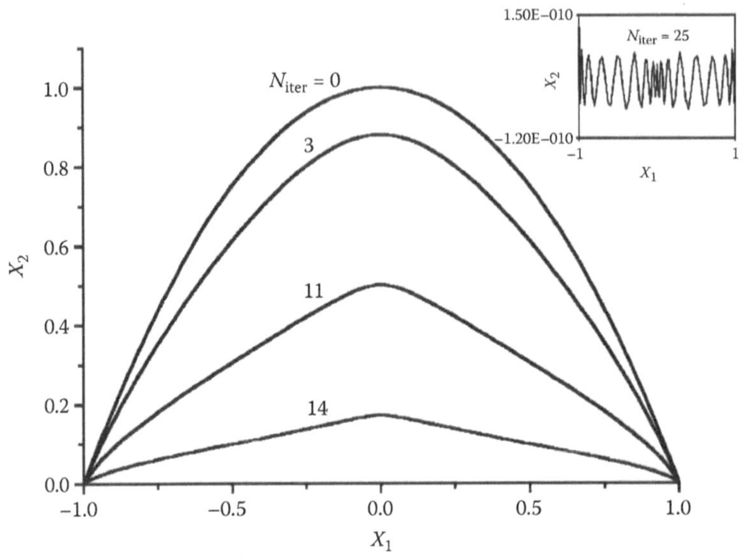

FIGURE 7.1 Iteration process of calculating the instanton trajectory in the separable potential model. N_{iter} is the number of iterations. The inset shows the result after 25 iterations. (Taken from Reference [43] with permission.)

trajectory. The initial guess is taken to be a straight line in $\mathbf{q}$−space connecting two potential minima, and 10 basis functions of the type of Equation (6.112) are used. For illustrative purposes, the back transformation $\mathbf{q} \to \mathbf{x}$ is made and the shapes of the trajectories in (x_1, x_2) plane are shown in Figure 7.1. The iteration converges to the straight line $x_i = 0 (i = 2, \ldots, N)$ and the final result was obtained after 25 iterations. The numerical error does not exceed 1.5×10^{-10} a.u., which allows us to obtain at least seven significant digits in the action W_0. The accuracy increases up to nine significant digits when 30 basis functions are used.

For solving the matrix differential equation, Equation (6.97), we make use of the analytical form of the instanton, Equation (6.111), and rewrite Equation (6.97) in terms of the parameter z as

$$\frac{d\tilde{\mathbf{A}}}{dz} = \frac{1}{\dot{z}(z)} R[\mathbf{A}, \mathbf{q}_0(z), \dot{\mathbf{q}}_0(z)], \tag{7.5}$$

where $R[\mathbf{A}, \mathbf{q}_0(z), \dot{\mathbf{q}}_0(z)]$ is the right-hand side of Equation (6.97). Since at $z = -1$ the time derivative of the parameter is zero, $\dot{z}(-1) = 0$, we actually have to integrate Equation (7.5) in the interval $[-1 + \epsilon_1, 0]$ where ϵ_1 is some small positive number. In so doing, we must first redefine the initial conditions at $z = -1 + \epsilon_1$. For this purpose, we implement the expansions

$$\mathbf{A}(z) = \mathbf{A}_m + \mathbf{a}(1 + z), \tag{7.6}$$

$$\dot{z}(z) = \alpha(1 + z), \tag{7.7}$$

$$H_{\mathbf{qq}}(z) = H_{\mathbf{qq}}(-1) + h_{\mathbf{qq}}(z + 1), \text{ etc.}, \tag{7.8}$$

where α and $h_{\mathbf{qq}}$ are known parameters: α is calculated from Equation (6.110). Equation (7.5) is solved near $z = -1$ by linear approximation, which gives the unknown matrix $\mathbf{a}$ and the initial condition for Equation (7.5) as $\mathbf{A}(-1+\epsilon_1) = \mathbf{A}_m + \epsilon_1 \mathbf{a}$. In turn, the second exponential factor W_1 can be rewritten as

$$W_1 = \frac{1}{2}\int_{-1}^{0} I(z)dz, \tag{7.9}$$

where

$$I(z) = \frac{1}{\dot{z}}\left[\mathrm{Tr}\left(\tilde{\mathbf{A}}(z) - \mathbf{A}_m\right) + \mathbf{p}^T \Lambda\right]. \tag{7.10}$$

The fourth-order Runge-Kutta method is used for solving Equation (7.5). In Figure 7.2 the numerical and analytical results are compared for $\epsilon_1 = 0.005$. Figure 7.2(a) presents the behavior of the diagonal element A_{11}. The accuracy is not worse than three significant digits in the whole interval. The smallness of $\dot{z}$ in the vicinity of $z = -1$ leads to rather poor accuracy of $I(z)$ in this region [see Figures 7.2(b) and 7.2(c)]. This error grows with the dimensionality of the system and seriously degrades the quality of the result. As mentioned above, however, the linear behavior of $\mathbf{A}(z)$ and $\dot{z}(z)$ at $z \simeq -1$ enables us to estimate $I(z)$ there accurately and we can avoid the numerical integration over the *bad* region by dividing the integral into two parts:

$$2W_1 = \left(\int_{-1}^{-1+\epsilon_2} + \int_{-1+\epsilon_2}^{0}\right) I(z)dz. \tag{7.11}$$

The first integral can be estimated as

$$\int_{-1}^{-1+\epsilon_2} \cdots dz = \frac{\epsilon_2}{2}\left(I(-1) + I(-1+\epsilon_2)\right). \tag{7.12}$$

This significantly improves the accuracy of W_1. We have used $\epsilon_2 = 0.025$ and 0.05, and confirmed three stable digits in the tunneling splitting $\Delta_0 = 0.624\ \mathrm{cm}^{-1}$. The analytical one-dimensional answer gives $\Delta_0 = 0.6238\ \mathrm{cm}^{-1}$.

7.2 HYDROPEROXY RADICAL HO_2

The second example is the hydrogen atom tunneling in the triatomic molecule HO_2 for zero total angular momentum ($J = 0$). In this case an analytical DMBE (double many-body expansion) potential function is available [126]. Since the quantum mechanically exact calculation can be carried out, the numerical results based on the present semi-classical theory are compared with the exact results. The exact numerical calculations are carried out in the hyperspherical elliptic coordinates (ρ, ξ, η) [127,128], which are convenient for describing a light atom transfer between two heavy atoms. The coordinate ρ is called hyperradius, which represents the mass scaled radius of the hypersphere and is defined as

$$\rho = \sqrt{\tilde{r}^2 + \tilde{R}^2}, \tag{7.13}$$

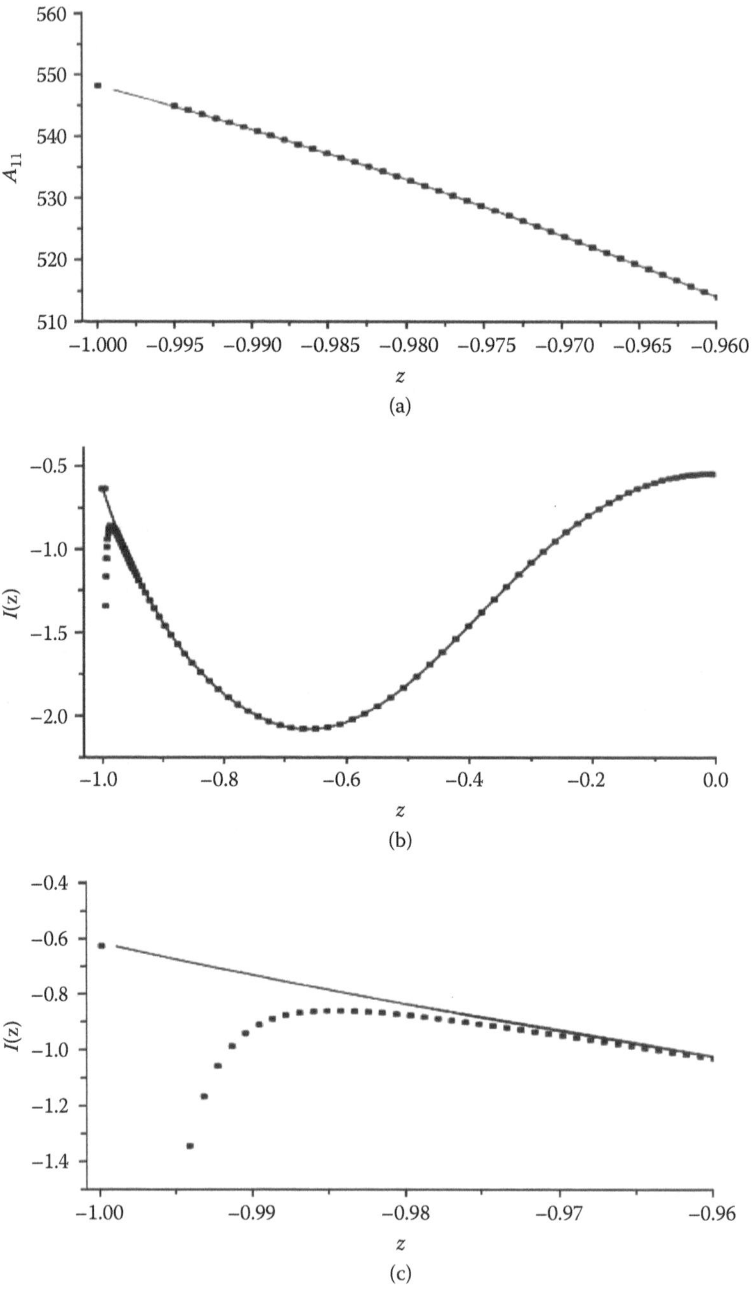

FIGURE 7.2 (a) Solution of the matrix equation, Equation (6.97), for the separable model. Only the diagonal element $\tilde{A}_{11}$ is shown. Solid squares: numerical results; solid line: exact analytical solution. (b) Behavior of the integrand $I(z)$. Solid squares: numerical solution; solid line: exact result. (c) Magnified portion of (b) to show the "bad" region in the vicinity of $z = -1$. ϵ_1 and ϵ_2 ($\epsilon_2 > \epsilon_1$) are taken to be 0.005 and 0.05, respectively [see the sentence below Equation (7.8) and Equation (7.11)]. (Taken from Reference [43] with permission.)

where $\tilde{r}$ and $\tilde{R}$ are the mass scaled Jacobi coordinates,

$$\tilde{r} = \sqrt{\frac{m_{O_2}}{\mu}} r \tag{7.14}$$

and

$$\tilde{R} = \sqrt{\frac{m_{H-O_2}}{\mu}} R. \tag{7.15}$$

Here, $r(R)$ is the distance between two oxygen atoms (hydrogen and the center of O_2), and the masses m_{O_2}, m_{H-O_2}, and μ are defined as

$$m_{O_2} = \frac{m_O}{2}, \quad m_{H-O_2} = \frac{2m_O m_H}{2m_O + m_H}, \text{ and } \mu = \sqrt{\frac{m_H m_O^2}{2m_O + m_H}}, \tag{7.16}$$

where m_I is the mass of atom I. The two hyperangles ξ and η are defined by

$$\xi = \chi_{O_1} + \chi_{O_2} \text{ and } \eta = \chi_{O_1} - \chi_{O_2}, \tag{7.17}$$

where χ_{O_j} is the Delves hyperangle and is given by

$$\chi_{O_j} = 2\arctan\frac{\tilde{r}_j}{\tilde{R}_j} \quad (j = 1, 2), \tag{7.18}$$

where $\tilde{r}_{1(2)}$ and $\tilde{R}_{1(2)}$ are the mass scaled Jacobi coordinates for the configuration $O_{1(2)} + HO_{2(1)}$. The $J = 0$ quantum mechanical Hamiltonian H takes the form

$$H = -\frac{1}{2\mu}\left[\frac{1}{\rho^5}\frac{\partial}{\partial\rho}\rho^5\frac{\partial}{\partial\rho} + \frac{1}{\rho^2}\Delta_{\xi\eta}\right] + V(\rho, \xi, \eta), \tag{7.19}$$

where $V(\rho, \xi, \eta)$ is the potential energy function and $\Delta_{\xi\eta}$ is the angular part of the kinetic energy operator defined by

$$\Delta_{\xi\eta} = \frac{16}{\rho^2(\cos\eta - \cos\xi)}\left[\frac{\partial}{\partial\eta}(\cos\eta - \cos 2\gamma)\frac{\partial}{\partial\eta} + \frac{\partial}{\partial\xi}(\cos 2\gamma - \cos\xi)\frac{\partial}{\partial\xi}\right], \tag{7.20}$$

where $\gamma \simeq 0.346$ is the angle of kinematic rotation [129] between the two Jacobi coordinates for the two different $O + HO$ arrangements and is given by

$$\tan\gamma = \sqrt{\frac{m_H(2m_O + m_H)}{m_O^2}}. \tag{7.21}$$

The ranges of the hyperspherical elliptic coordinates span as follows:

$$0 < \rho < \infty, \quad -2\gamma < \eta < 2\gamma, \text{ and } 2\gamma < \xi < 2\pi - 2\gamma. \tag{7.22}$$

The hyperangle η roughly represents the rotational motion of hydrogen around two oxygen atoms and the potential energy function is symmetric with respect to η inversion. Further details about the hyperspherical Delves and elliptic coordinate systems can be found in References [127,128].

The height of the potential barrier separating the two equivalent potential wells is as high as ~ 2.4 eV; as a result, the energy difference between the *gerade* and *ungerade* energy levels of lower excited states appears only in the 11th to 12th significant digit, which is far beyond the accuracy of the quantum calculations. To overcome this difficulty, we utilize an analog of the Herring formula written in terms of *exact* wave functions. Using Green's theorem, the energy difference between two levels can always be expressed as a surface integral of some bilinear combination of the corresponding two wave functions. It is confirmed that the calculations of the quantum wave functions secure at least two to three significant digits in the area of a potential barrier so that the energy splitting can be evaluated with the same accuracy. To derive the formula of energy splitting $\Delta_{12} = E_1 - E_2$, we consider the Schrödinger equation for the corresponding eigenstates Ψ_1 and Ψ_2. As usual, multiplying the equation for $\Psi_1(\Psi_2)$ by $\Psi_2(\Psi_1)$, subtracting one from the other, and integrating with the weight $\rho^2(\cos\eta - \cos\xi)$ over the region $\rho \in [0, \infty]$, $\xi \in [2\gamma, 2\pi - 2\gamma]$, $\eta \in [-2\gamma, 0]$, we obtain

$$\Delta_{12} = \frac{8(1-\cos 2\gamma)\int_0^\infty d\rho \int_{2\gamma}^{2\pi-2\gamma} d\xi \left(\Psi_1 \frac{\partial\Psi_2}{\partial\eta} - \Psi_2 \frac{\partial\Psi_1}{\partial\eta}\right)_{\eta=0}}{\int_0^\infty \rho^2 d\rho \int_{2\gamma}^{2\pi-2\gamma} d\xi \int_{-2\gamma}^{0} d\eta(\cos\eta - \cos\xi)\Psi_1\Psi_2}. \tag{7.23}$$

This is valid for any two quantum states. Now we consider the wave functions $\Psi_{1,2} = \Psi_{+,-}$ of the pair of states having opposite symmetries with respect to η inversion. The second term in the numerator vanishes, because $\Psi_2(\eta = 0) = 0$ and for low energy levels the normalization integral can be substituted by 1/2, which gives the desired formula

$$\Delta = 16(1-\cos 2\gamma)\int_0^\infty d\rho \int_{2\gamma}^{2\pi-2\gamma} d\xi\, \Psi_+ \frac{\partial\Psi_-}{\partial\eta}|_{\eta=0}. \tag{7.24}$$

Equation (7.24) is used to evaluate the exact value of the tunneling splitting. We seek the solution of the Schrödinger equation in the form of SVD (smooth variable discretization) expansion, [130], which is essentially the same as the sequential truncation diagonalization technique [131],

$$\Psi_\pm = \sum_{i=1}^{N_{DVR}} \sum_{\upsilon=1}^{N_{ad}} C_{i\upsilon}\tilde{\pi}_i(\rho)\Phi_\upsilon^\pm(\xi, \eta; \rho_i), \tag{7.25}$$

where $\Phi_\upsilon^\pm(\xi, \eta; \rho)$ are the adiabatic channel functions, even (+) or odd (−) with respect to η inversion; i.e., the eigenfunctions of the two-dimensional Hamiltonian on the hypersphere, $(1/2\mu\rho^2)\Delta_{\xi\eta} + V(\rho, \xi, \eta)$, at fixed value of hyperradius ρ, $\tilde{\pi}_i(\rho)$ are the discrete variable representation (DVR) basis functions, and ρ_i are the corresponding DVR quadrature points. At each quadrature point the adiabatic channel functions were calculated in the same way as that used before [132]. The whole interval of hyperradius $[\rho_0, \rho_f]$ propagated in the numerical calculation is divided into a number of sectors and in each sector the above expansion, Equation (7.25), is made. The DVR functions are constructed from Laguerre polynomials $L_n^{(2)}[\alpha(\rho - \rho_0)]$, where the scaling parameter $\alpha = X_{N_{DVR}}/(\rho_f - \rho_0)$ is estimated from the last quadrature point $X_{N_{DVR}}$ and the interval $[\rho_0, \rho_f]$ where the wave functions of the bound states are localized.

TABLE 7.1
Bound State Energies (in eV) of HO_2 Radical for Total Angular Momentum $J = 0$ and Odd O-O Exchange Parity Relative to the Ground-State Energy. The Assignment of the Levels Is Taken from [133]. n_1, n_2, and n_3 Are the Quantum Numbers Corresponding to HO Stretch, HOO Bend, and O_2 Stretch, Respectively

n	(n_1, n_2, n_3)	E: Reference [133]	E: present
1	(0, 0, 0)	0	0
2	(0, 0, 1)	0.13210	0.13209
3	(0, 1, 0)	0.16073	0.16072
4	(0, 0, 2)	0.25927	0.25923
5	(0, 1, 1)	0.29254	0.29252
6	(0, 2, 0)	0.31203	0.31201
7	(0, 0, 3)	0.38200	0.38195
8	(1, 0, 0)	0.41332	0.41330
9	(0, 1, 2)	0.41993	0.41989
10	(0, 2, 1)	0.44508	0.44504
11	(0, 3, 0)	0.46194	0.46191
12	(0, 0, 4)	0.50105	0.50099
13	(0, 1, 3)	0.54163	0.54157
14	(1, 0, 1)	0.54819	0.54814
15	(0, 2, 2)	0.57226	0.57221
16	(1, 1, 0)	0.58128	0.58124
17	(0, 3, 1)	0.59724	0.59720
18	(0, 4, 0)	0.61303	0.61301
19	(0, 0, 5)	0.61693	0.61686
20	(0, 1, 4)	0.66115	0.66108
21	(1, 0, 2)	0.67545	0.67540
22	(0, 2, 3)	0.69563	0.69557
23	(1, 1, 1)	0.71204	0.71200
24	(0, 3, 2)	0.72543	0.72538
25	(0, 0, 6)	0.73009	0.73001
26	(1, 2, 0)	0.73298	0.73294
27	(0, 4, 1)	0.75036	0.75032
28	(0, 5, 0)	0.76318	0.76315
29	(0, 1, 5)	0.77742	0.77734
30	(1, 0, 3)	0.79609	0.79603
31	(2, 0, 0)	0.80494	0.80490
32	(0, 2, 4)	0.81542	0.81535

Further details about the DVR method can be found in Reference [132]. In the present calculations $\rho_0 = 3.2$ and $\rho_f = 8.5$ (in atomic units) are used, which enables us to obtain enough accuracy for the lowest 300 bound states. The spectral calculation has been performed separately for even and odd parity states. The size of the DVR basis and the number of the adiabatic surface functions were taken to be $N_{DVR} = 130$ and $N_{ad} = 15$. To check the accuracy the calculations with $N_{DVR} = 140$ and $N_{ad} = 25$ were repeated. Table 7.1 shows the calculated bound state energies in comparison with the results of Reference [133] obtained by the filter diagonalization method. The disagreement in the last digit is probably due to the fact that the subroutines for DMBE potential energy surface were not completely identical. For the lowest 50 levels, the three stable digits in the wave function on the dividing surface $\eta = 0$ are confirmed,

TABLE 7.2
Tunneling Splitting of the Low Vibrationally Excited States in HO_2 Radical for Total Angular Momentum $J = 0$ (see Table 7.1). The Exact Value of the Ground-State Tunneling Splitting Is Δ_0(exact) $= 0.77 \times 10^{-12}$ eV and its Corresponding Semiclassical Value Is Δ_0(semicl.) $= 0.78 \times 10^{-12}$ eV (see text)

n	(n_1, n_2, n_3)	Δ_n/Δ_0(exact)	Δ_n/Δ_0(semicl.)
1	(0, 0, 0)	1.0	1.0
2	(0, 0, 1)	1.82	1.62
3	(0, 1, 0)	0.77×10^2	1.02×10^2
5	(0, 1, 1)	1.34×10^2	1.65×10^2
8	(1, 0, 0)	1.12×10^3	0.9×10^3
14	(1, 0, 1)	2.32×10^3	2.0×10^3
16	(1, 1, 0)	5.76×10^4	8.7×10^4

which makes it possible to use Equation (7.24). The calculated values of tunneling splittings are shown in Table 7.2 in comparison with the semiclassical results. The absolute value of the ground-state splitting is $\Delta_0(\text{exact}) = 0.77 \times 10^{-12}$ eV.

In the semiclassical approximation, one can express $\Psi_\pm$ in terms of the wave functions Ψ_L and Ψ_R localized in each potential well as

$$\Psi_\pm = \frac{\Psi_L \pm \Psi_R}{\sqrt{2}}. \tag{7.26}$$

With this substitution, Equation (7.24) reduces to the Herring formula for the particular case of the hyperspherical elliptic coordinates in triatomic system with $f = \eta$. This semiclassical calculation of the ground-state tunneling splitting is performed by using the hyperspherical elliptic coordinates and Figure 7.3 demonstrates the convergence of the instanton trajectory for the projection of the system onto the hypersphere $\rho = 4.48$ a.u. [43]. The ground-state tunneling splitting obtained is $\Delta_0(\text{semicl.}) = 0.78 \times 10^{-12}$ eV, which is in good agreement with the exact value mentioned above.

The semiclassical calculations for the excited states are carried out by using the methods explained in Sections 6.1 and 6.4 with use of the internal shape space coordinates instead of the hyperspherical elliptic coordinates. The instanton trajectory was calculated by using 20 and 30 Legendre basis functions and the results are found to be identical up to six significant digits in the classical action. The matrix solution $\tilde{\mathbf{A}}(z)$ of Equation (6.181) is used for the excited-state calculations. There are three vibrational modes in the system with the corresponding normal frequencies, $\omega_\gamma = 0.1364$, 0.1677, and 0.4320 (all in eV). The difference between ω_γ and the exact value ΔE_γ (see $n = 2, 3, 8$ of Table 7.1) indicates the strong anharmonicity effect that is most pronounced in the highest frequency mode. The quantum calculation shows a very strong state selectivity such that for different excited modes, γ, the ratio $\Delta_{n_\gamma=1}/\Delta_0$ differs in orders of magnitude (see Table 7.2).

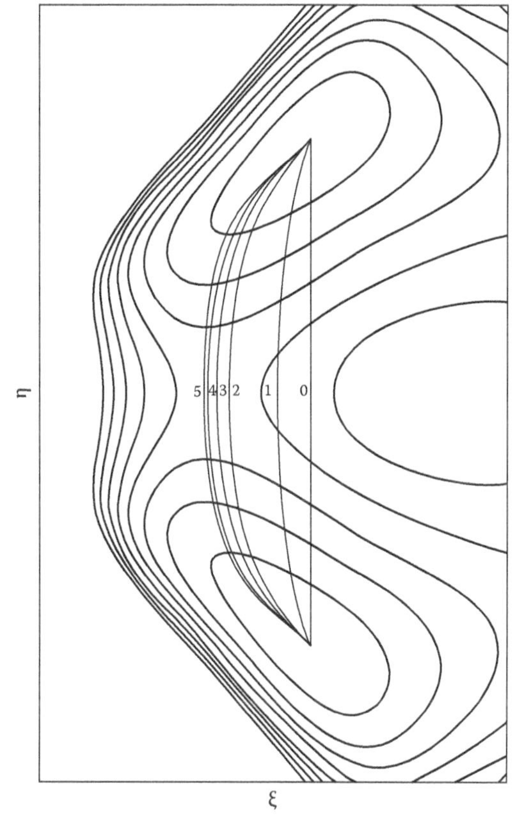

N_{iter}	$2W_{\text{OE}}$
0	40.512
1	36.699
2	31.496
3	29.606
4	28.832
5	28.752
6	28.752

FIGURE 7.3 Iteration process for the calculation of instanton trajectory in the case of HO_2. The projection onto the hypersphere at $\rho = 4.48$ a.u. is shown. η and ξ are the two elliptic hyperangles. N_{iter} and $2W_{0\Sigma}$ are the number of the iteration and the corresponding classical action, respectively. (Taken from Reference [43] with permission.)

The ratio $\Delta_{n_\gamma=1}/\Delta_0$ is given by Equation (6.175) for the transversal modes and by Equation (6.140) for the longitudinal mode. The latter can be identified either from the time-dependence of the parameter $z(\tau)$, i.e., $\dot{z}/(1+z)|_{z=-1} = \omega_{\parallel}$, or from the direction of the velocity at the potential minimum. We obtained $\omega_{\parallel} = 0.1675$ eV and $[(d\mathbf{q}_0/dz)\cdot\mathbf{n}_2]_{z=-1} \simeq 0.998|d\mathbf{q}_0/dz|\cdot|\mathbf{n_2}||_{z=-1}$, where $\mathbf{n}_2$ is the corresponding direction for the second normal mode with $\omega_\gamma = 0.1677$. Thus we treat the excitation of the mode 2 ($n_2 = 1$) as the longitudinal one and use Equation (6.140), while for the modes 1 ($n_3 = 1$) and 3 ($n_1 = 1$) (see Table 7.1) we apply Equation (6.175). In the first case, straightforward calculations give $\Delta_{n_\gamma=1}/\Delta_0 = 102$, which coincides with the quantum result within $\sim 25\%$ (see Table 7.2). The calculation of the tunneling splitting for the transversal excitations is less trivial, since one has to deal with Equation (6.191). By making the transformation of the independent variable $\tau \to z$, we bring it into the form

$$\dot{z}(z)\frac{dU_k}{dz} = \theta(z)U_k - \sum_{ij}\left[g^{ij}\tilde{A}_{ik} + \frac{\partial g^{ij}p_{0i}}{\partial q^k}\right]U_j. \tag{7.27}$$

All the terms in this equation are analytical functions of the parameter z except for $\tilde{\mathbf{A}}(z)$, which is the numerical solution of Equation (6.181). Equation (7.27) is a system of coupled nonlinear equations with removable singularity at $z = -1$ that may cause instability of the numerical solution. Note, however, that the nonlinearity of this equation is totally due to the factor $\theta(z)$ [see Equation (6.193)], which is related to the normalization condition for the vector $\mathbf{U}$. Thus, for a given mode γ the solution can be written in the form

$$\mathbf{U} = \frac{\tilde{\mathbf{U}}}{\sqrt{\tilde{\mathbf{U}}\mathbf{g}\tilde{\mathbf{U}}}}, \tag{7.28}$$

where $\tilde{\mathbf{U}}$ satisfies a system of *linear* equations

$$\dot{z}(z)\frac{d\tilde{U}_k}{dz} = \omega_\gamma \tilde{U}_k - \sum_{ij}\left[g^{ij}\tilde{A}_{ik} + \frac{\partial g^{ij} p_{0i}}{\partial q^k}\right]\tilde{U}_j. \tag{7.29}$$

To avoid the problem of "stiffness" [134] in the vicinity of the singular point $z = -1$, we accurately fit all the terms in Equation (7.29) within a small initial interval $z \in [-1, -1+\epsilon]$ by finite-order polynomial functions of $(z+1)$ and then explicitly find a solution in the *infinite* order expansion

$$\mathbf{U}(z) = \sum_{n=0}^{\infty} \mathbf{C}_n (1+z)^n, \tag{7.30}$$

where $\mathbf{C}_0$ is the eigenvector of the matrix $\mathbf{Ag}$ with the eigenvalue ω_γ and the other coefficients $\mathbf{C}_n (n > 0)$ can be found from the recurrence relations that follow from Equation (7.29). The desired solution $\mathbf{U}(-1+\epsilon)$ is obtained by the normalization and Equation (7.29) can be further propagated until $z = 0$ by any conventional numerical method. Once this is done for both transversal modes, the tunneling splitting of the low excited states can be computed. The semiclassical calculations are very fast compared with the exact quantum calculations. The final results are presented in Table 7.2. As expected, the accuracy of the semiclassical results for the excited states is worse as compared with the ground-state case. From the numerical analysis, however, it has turned out that the $\sim$ 20% of numerical inaccuracy of the results mainly comes from the inaccuracy of the normalization factor, i.e., in the denominator of Equation (6.195). In principle, the Hamilton-Jacobi equation can be solved in the region of the potential minimum more accurately. This needs a further analysis.

As can be seen in Table 7.2, the peculiarity in the effect of each vibrational mode is perfectly reproduced by the modified WKB theory. It is interesting to clarify the origin of almost three orders of magnitudes difference in the tunneling splitting between two transversal excitations. The main factor in the semiclassical formula that dictates the tunneling splitting growth is the behavior of the effective frequency $\theta(\tau)$. In Figure 7.4 we compare the computed $\theta(\tau)$ with the adiabatic approximation $\theta_{ad}(\tau)$ given by the corresponding eigenvalue of the matrix $\mathbf{g}\tilde{\mathbf{A}}$. $\theta(\tau)$ has the meaning of local adiabatic transversal frequencies along the instanton, which are often used in simplified versions of the instanton approach to describe the excitations. For the low energy excitation, $\theta(\tau)$ is close to $\theta_{ad}(\tau)$ and one obtains a moderate growth of Δ_{n_γ} similar to the adiabatic approximation. On the contrary, for the high energy

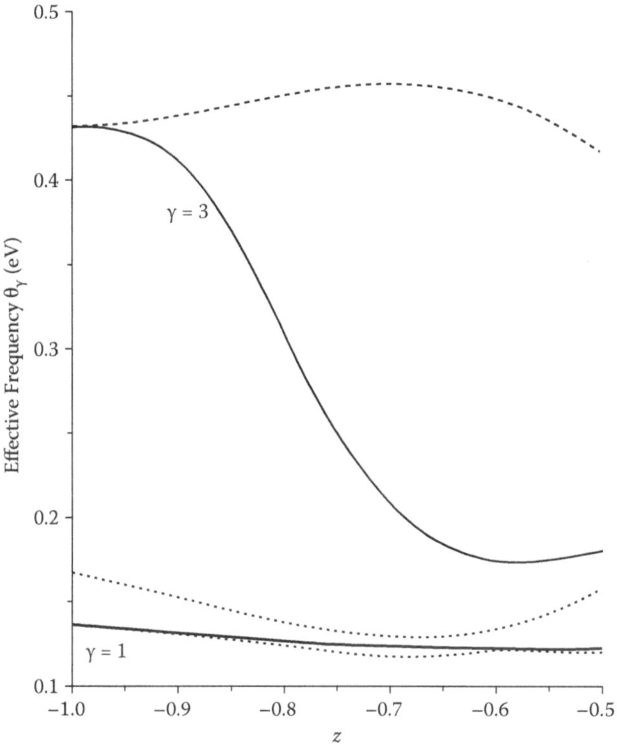

FIGURE 7.4 The effective frequency $\theta(\tau)$ in the integrand of Equation (6.170) for the two transversal modes $\gamma = 1(n_3 = 1), 3(n_1 = 1)$ in HO_2 radical. Solid lines show the results of numerical solutions of Equations (6.191) and (6.193). Dotted lines represent the adiabatic frequencies. (Taken from Reference [45] with permission.)

excitation, $\theta(\tau)$ rapidly decreases due to the interaction with the low frequency mode, which leads to large negative values of ΔS_1 in Equation (6.170) and exponential growth of the tunneling splitting.

As was pointed out before, such a dramatic effect of the vibrational excitation cannot be observed in two-dimensional models, since in that case $\theta(\tau)$ just coincides with $\theta_{ad}(\tau)$. The analysis of Equation (7.29) in the vicinity of $z = -1$ shows that for a certain γ_0th excited normal mode the recurrence relations between the coefficients in the expansion Equation (7.30) contains the inverse of the matrix with the eigenvalues $(\omega_{\gamma_0} - \omega_\gamma - n\omega_\parallel)$. If for some n and γ, $\omega_{\gamma_0} \simeq \omega_\gamma + n\omega_\parallel$ happens, the radius of convergence of Equation (7.30) becomes very small, which leads to a very sharp change of $\theta(z)$. It can be shown that since $\omega_\gamma < \omega_{\gamma_0}$, $\theta(z)$ decreases so that the integrand $(\theta - \omega_{\gamma_0})/\dot{z}$ may take negative large values. Note that in the extreme case of the exact resonance, one would get $\Delta S_1 \to \infty$. This, of course, indicates a breakdown of the instanton theory, since, at the very least, one expects that the condition $\Delta S_1 \ll S_0$ holds.

7.3 VINYL RADICAL C_2H_3

The vinyl radical (see Figure 7.5) is an important intermediate in combustion chemistry, attracting much interest for many years [135,136]. Intensive works on its structure have been carried out both theoretically [137–141] and experimentally [142–149]. In particular, Fessenden and Schuler observed a pair of doublets ascribed to an α proton in the CH group and two β protons in the CH_2 group [144]. They concluded that the absence of central $\alpha\beta$ and $\beta\alpha$ lines is due to the fast interconversion of the C–H bond between two minima and estimated the corresponding potential barrier height to be about 2 kcal/mol. Kanamori et al. reported the results obtained by the IR diode laser kinetic spectroscopy [148]. They observed the splitting of an absorption band around 900 cm^{-1} assigned to the out-of-plane CH_2 wagging motion. Although the tunneling splitting in the ground state Δ_0 cannot be directly estimated from their data, the authors derived the potential barrier height 1200 cm^{-1} from the analysis of rotational constants. Quite recently, Tanaka et al. [149] investigated this radical by millimeter-wave spectroscopy and reported a set of precise molecular constants. Among them the ground-state tunneling splitting was found to be Δ_0 =16272 MHZ (= 0.54 cm^{-1}). Using a one-dimensional double well model, they also estimated the barrier height as 1580 cm^{-1} as well as the tunneling splitting of the rotational constant. In addition to the tunneling splitting, here we estimate the splitting of rotational constants by using the semiclassical wave function [150]. The vinyl radical is a good example in the sense that the full-scale semiclassical calculations with use of a high level of quantum chemical calculations can be accomplished. We can asses the accuracy of the semiclassical theory.

As was explained in Section 6.2, the body-fixed frame of reference is specified by imposing six conditions on the 15 body-fixed Cartesian coordinates $\mathbf{r}_n = \sum_{i=1}^{3} x_{in}\mathbf{e}_i (n = 1, 2, \ldots, 5)$. The first three conditions fix the origin to the center of mass. The other three conditions specify the orientation of the body-fixed axes

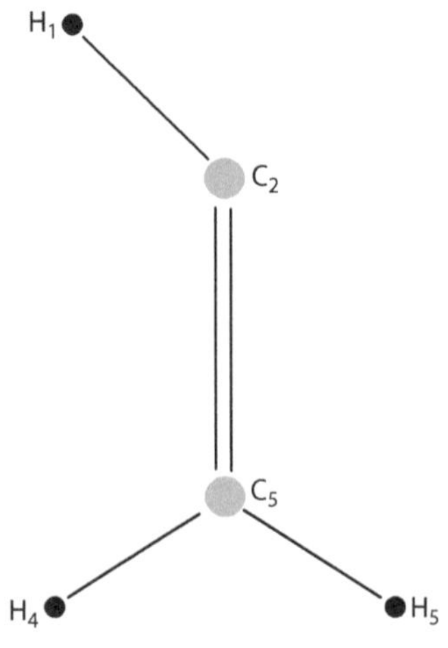

FIGURE 7.5 Vinyl radical. (Taken from Reference [150] with permission.)

TABLE 7.3
Normal Frequencies (in cm^{-1}) of C_2H_3

N	type of motion	ω_γ CCSD(T)/aug-cc-pVTZ	MP2/6-31G(d,p)
1	C_2H_1rocking vibration	711	771
2	wagging(out of plane)	813	996
3	wagging(out of plane)	923	1063
4	plane distortion	1062	1129
5	$H_4C_3H_5$bending	1390	1465
6	C_2C_3stretching	1632	1863
7	H_5C_3stretching	3065	3203
8	$H_1H_2H_3$assym.bending	3171	3305
9	C_2H_1stretching	3238	3360

$(\mathbf{e}_1, \mathbf{e}_2, \mathbf{e}_3)$ in such a way that the tunneling hydrogen H_5 lies in the $(\mathbf{e}_1, \mathbf{e}_2)$ plane and $\mathbf{e}_1$ is directed along the line connecting the two hydrogen atoms in the CH_2 group. With use of these conditions, the metric tensor for the rotation-free ($J = 0$) quantum Hamiltonian is constructed as explained in Chapter 6. In the iterative procedure to find instanton trajectory, we used 12 reference points at each step. The quantum chemical calculations were performed at the levels of MP2/6-31G(d,p) [151,152] and of CCSD(T)/aug-cc-pVTZ [124,153] of electronic structure theory. The gradients and Hessians required at each point were calculated analytically at the MP2 level and numerically at the CCSD(T) level using the GAUSSIAN program [154].

Table 7.3 shows a comparison of the normal mode frequencies by the two methods (third and fourth columns). The instanton trajectory was first calculated at the MP2 level, where computations are quick and analytical second derivatives are available. Starting from the straight line connecting the two potential minima as the initial trial path, full convergence was achieved after 10 iterations with five stable significant digits in the classical action guaranteed. With use of this instanton trajectory, Equation (6.97) was solved by the standard Runge-Kutta method and the tunneling splitting of the ground state was calculated. At the MP2/6-31G(d,p) level, we obtained $\Delta_0 = 0.14\ cm^{-1}$, which is about four times smaller than the experimental value (see Table 7.4). This discrepancy is likely to be ascribed to the insufficient accuracy of the potential data. In particular, the potential barrier along the instanton path at this MP2 level turns out to be as high as 2249 cm^{-1}. Our experience indicates that one cannot generally expect any accurate results from the simulations on the MP2 potential energy surface. At the same time, this preliminary step is necessary in order to reduce the numerical efforts at the CCSD(T)/aug-cc-pVTZ level. This is an important feature of our iterative method, which enables us to use the obtained instanton path as the initial trial guess for the higher level of computations. The choice of the *ab initio* methods at the preliminary stage is not important, and any numerically cheap method can be used. Due to the similarity of potential topology, usage of such a trial path reduces

TABLE 7.4
Tunneling Splitting (cm^{-1}) of the Ground-State Energy Levels of Vinyl Radical

method	classical barrier[a]	effective barrier[b]	Δ_0	B'	S_0	S_1
MP2/6-31G(d, p)	2233	2249	0.14	2234	10.65	−0.98
CCSD(T)/aug-cc-pVTZ	1761	1770	0.53	2023	9.37	−1.11
experiment		1580[c,d]	0.54[c]			

[a]The difference between the saddle point and the minimum without zero-point energy correction.
[b]The barrier height along the instanton path.
[c]Reference [149].
[d]One-dimensional model is used.

the number of iterations at the higher level and the calculation can be completed within a reasonable time effort. In the present case, only two extra iterations turn out to be enough to obtain three stable significant digits in the classical action at the CCSD(T)/aug-cc-pVTZ level of electronic structure calculations. The convergence of the iterative procedure is illustrated in Figure 7.6, which shows the projection of the instanton path onto the (X, Y) plane for the tunneling hydrogen atom. The paths numerated from 1 to 9 correspond to the successive steps of iteration using the MP2 *ab initio* calculations. As can be seen from Figure 7.6, the ninth path and the path at the CCSD(T)/aug-cc-pVTZ level are essentially the same in *shape* and the main difference consists of the shift in the minimum positions. This is a common feature observed in other systems.

At the CCSD(T)/aug-cc-pVTZ level, our theory gives the ground-state splitting as $\Delta_0(\text{theor.}) = 0.53\ \text{cm}^{-1}$, which almost perfectly reproduces the experimental value $\Delta_0(\text{exp.}) = 0.54\ \text{cm}^{-1}$ (see Table 7.4). In this table also shown are the three factors $B' = B\exp[-S_1]$, S_0, and S_1, when we write $\Delta_0 = B'\exp(-S_0/\hbar - S_1)$ [see, for instance, Equation (6.100) and (6.117), where $S_0 = 2W_{0\Sigma}$ and $S_1 = 2W_{1\Sigma}$]. As one can see, the difference in Δ_0 between the two methods comes mainly from the principal exponent S_0 and the other two factors are fairly close to each other. This also confirms the above surmise about the similarity in the potential topology for different *ab initio* methods. On the other hand, the principal exponent S_0 is mainly affected by the height of the potential barrier, which in the case of CCSD(T)/aug-cc-pVTZ reduces to 1770 cm^{-1}. Note that there is almost 200 cm^{-1} difference from the estimate made by Tanaka et al. [149], which is probably due to the drawback of their one-dimensional model.

Considering that the accuracy of the quantum chemical calculations is most crucial, the transition state and effective barrier height energies are calculated with use of various basis functions (see Table 7.5). Table 7.5 shows that the CCSD(T) barrier heights are almost the same as the corresponding MP2 ones with 6-31G(d,p), aug-cc-pVDZ, and aug-cc-pVTZ basis sets. One can further see that the MP2/aug-cc-pVQZ barrier height, 4.98 kcal/mol, is very close to 5.00 kcal/mol of MP2/aug-cc-pVTZ.

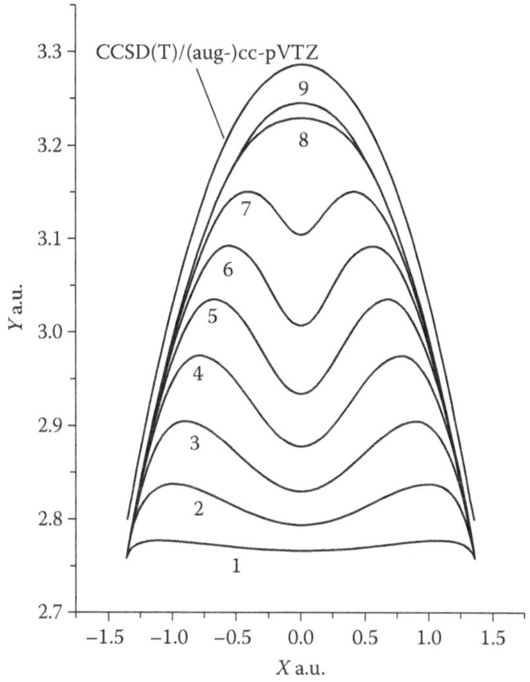

FIGURE 7.6 Iterative calculation of the instanton path in vinyl radical. The labeled paths from 1 to 9 show gradual improvement of the instanton trajectory shape using the MP2/cc-pVDZ *ab initio* data. After switching to the CCSD(T)/aug-cc-pVTZ *ab initio* method, only two more steps are required to achieve convergence and obtain the final result. (Taken from Reference [150] with permission.)

This tendency suggests that the CCSD(T)/aug-cc-pVQZ barrier height must be close to the one obtained by the CCSD(T)/aug-cc-pVTZ and this basis is large enough to provide a converged result. The same conclusion follows from the values of the effective barrier V_0 shown in the last column of Table 7.5. Assuming $S_0 \propto (V_0)^{1/2}$, we can estimate the accuracy of the classical action $\delta S_0 < 0.03$, which corresponds to $\sim 3\%$ error in the tunneling splitting. This indicates that the present semiclassical theory is quite accurate, if the quantum chemical *ab initio* calculations are carried out at a high level such as the present one.

We have applied the semiclassical theory of low excited states presented in Section 6.4 to nine normal-mode excitations. The results for $N = 1$–6 in Table 7.3 with use of the CCSD(T)/aug-cc-pVTZ and MP2/6-31G(d,p) methods are shown in Table 7.6. The longitudinal mode is the lowest one that corresponds to the rocking vibration of the tunneling hydrogen atom. This has also been checked numerically by calculating the projection $(d\mathbf{q}_0/d\tau)\mathbf{n} \simeq 0.99|(d\mathbf{q}_0/d\tau)||\mathbf{n}|$ at the potential minimum where $\mathbf{n}$ is the corresponding direction for the first normal mode. For this rocking mode excitation, the theory predicts about 40 times growth of the tunneling splitting and both of the *ab initio* methods give similar results. The situation changes in the case of transversal

TABLE 7.5
Barrier Height in the Vinyl Radical for Various *ab initio* Methods

ab initio Method	Classical Barrier[a]	Classical Barrier[b]	Effective Barrier[a,c]
MP2/6-31G(d, p)	2232.8	6.38	
MP2/aug-cc-pVDZ	2063.1	5.90	
MP2/aug-cc-pVTZ	1749.2	5.00	
MP2/aug-cc-pVQZ	1740.2	4.98	
CCSD/6-31G(d, p)	2224.2	6.36	
CCSD/aug-cc-pVDZ	2086.8	5.97	
CCSD/aug-cc-pVTZ	1805.5	5.16	
CCSD(T)/6-31G(d, p)	2209.7	6.32	
CCSD(T)/aug-cc-pVDZ	2058.5	5.89	
CCSD(T)/aug-cc-pVTZ	1761.3	5.04	1773.2
CCSD(T)/cc-pVQZ			1784.7
CCSD(T)/aug-cc-pVQZ			1768.1

[a]in cm^{-1}.
[b]in kcal/mol.
[c]The energy difference between the middle point of the CCSD(T)/aug-cc-pVTZ instanton path and the minimum.

mode excitation. For low energy modes, one observes a moderate growth of the tunneling splitting, but the result is more sensitive to the details of the potential energy surface compared with the rocking mode. For the fifth and sixth vibrational mode, one can expect a strong promotion effect and accurate evaluation of the potential energy surface is required. In Figure 7.7 we depict the effective frequency θ in Equation (6.193) as a function of the parameter z. The point $z = -1$ corresponds to $\tau = -\infty$, that is, the potential minimum. For all the transversal modes, the effective

TABLE 7.6
Tunneling Splitting for Excited States of Vinyl Radical

		Δ/Δ_0	
N	type of motion	CCSD(T)/aug-cc-pVTZ	MP2/6-31G(d,p)
1	C_2H_1rocking vibration	36.0	41.1
2	wagging(out of plane)	2.20	1.76
3	wagging(out of plane)	1.28	1.12
4	plane distortion	3.0	2.3
5	$H_4C_3H_5$bending	1.7×10	1.5
6	C_2C_3stretching	2×10^2	7×10

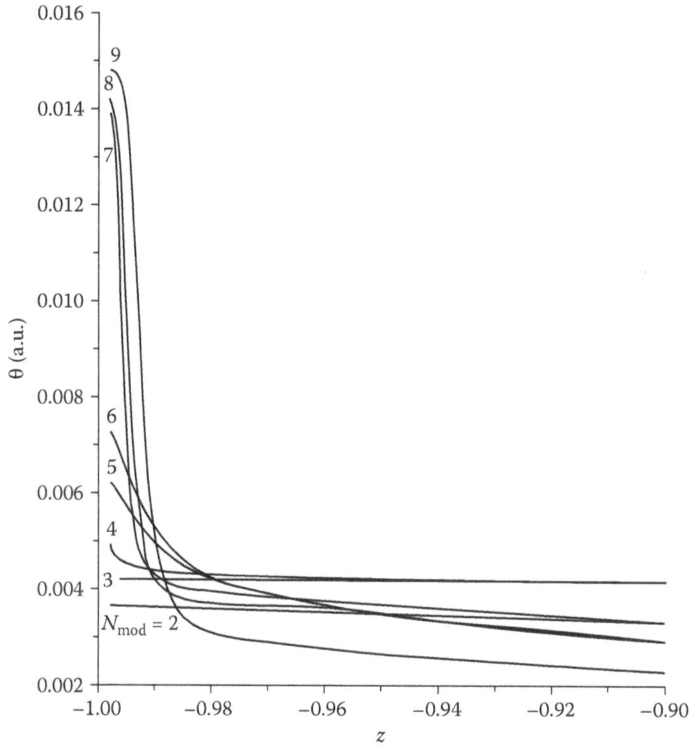

FIGURE 7.7 Effective frequency $\theta(z)$ in the integrand of Equation (6.170) for eight transversal excitations in the case of vinyl radical. The steep decrease of $\theta(z)$ in the case of high excitations indicates a strong interaction with low energy modes. This leads to the breakdown of the semiclassical theory for the modes $N = 7$–9. (Taken from Reference [150] with permission.)

frequency monotonically decreases, which leads to the negative exponential factor ΔS_1 in Equation (6.170). For higher frequency modes, the curve of θ is steeper, indicating a strong interaction with the lower frequency vibrations. In this region, the exponential factor S_0 becomes dominant, but more sensitive to the *ab initio* method. For the last three modes ($N = 7$–9) in Table 7.3, ΔS_1 becomes comparable with the principal exponential factor S_0, which indicates a breakdown of the semiclassical approximation. For the modes $N > 6$, the excitation energy exceeds the potential barrier and the simple picture of the semiclassical approximation does not work.

Before closing this section, we discuss a more general problem of the semiclassical estimation of physical quantity. The *ab initio* data obtained so far can be used to calculate more detailed information about the system, such as the splitting of rotational constant. For simplicity, we restrict our attention to the ground state with wave function given explicitly by Equations (6.130), (6.180), and (6.187). The calculation of the matrix element generally requires the global wave function, while Equation (6.180) gives only local behavior in close vicinity of the instanton trajectory. However, since our aim here is to estimate the difference between the expectation values of the two states of opposite symmetry in the tunneling doublet, the present semiclassical theory can be used.

We consider the expectation values

$$\mathbf{B}^{\pm} \equiv \frac{1}{2}\langle\Psi^{\pm}|\mathbf{G}|\Psi^{\pm}\rangle, \tag{7.31}$$

where $G^{kl}(\mathbf{q})(k, l = 1, 2, 3)$ is the rotational metric tensor, and $\Psi^{+}(\mathbf{q})$ and $\Psi^{-}(\mathbf{q})$ are the wave functions for symmetric and antisymmetric states in the ground-state doublet, respectively. We introduce the functions $\Psi_{1,2} = (\Psi^{+} \pm \Psi^{-})/\sqrt{2}$ localized in each potential well and rewrite Equation (7.31) in the form

$$\mathbf{B}^{\pm} = \mathbf{M} \pm \mathbf{m} \tag{7.32}$$

with

$$\mathbf{M} = \frac{1}{4}\Big(\langle\Psi_1|\mathbf{G}|\Psi_1\rangle + \langle\Psi_2|\mathbf{G}|\Psi_2\rangle\Big) \tag{7.33}$$

and

$$\mathbf{m} = \frac{\langle\Psi_1|\mathbf{G}|\Psi_2\rangle}{2}. \tag{7.34}$$

The rotational constants are defined as the eigenvalues of $\mathbf{B}^{\pm}$. By treating the second exponentially small term in Equation (7.32) as a perturbation, we obtain the tunneling splitting of rotational constant in the form

$$\Delta B_k = 2\mathbf{X}_k^T \mathbf{m} \mathbf{X}_k, \tag{7.35}$$

where $\mathbf{X}_k (k = 1, 2, 3)$ are the eigenvectors of $\mathbf{M}$. Within the limits of semiclassical accuracy, the latter must be taken in zeroth-order approximation, that is, $\mathbf{M} = \mathbf{G}(\mathbf{q}_m)/2$. The problem therefore is reduced to calculating the exponentially small term of Equation (7.34). Let us consider a matrix element

$$I = \langle\Psi_1|f|\Psi_2\rangle, \tag{7.36}$$

where $f(\mathbf{q})$ is an arbitrary function of coordinates that is assumed to possess appropriate symmetry. In the semiclassical approximation, the main contribution to the matrix element comes from the characteristic that is common for the families of classical trajectories associated with the functions Ψ_1 and Ψ_2. This is nothing but the instanton trajectory, and the matrix elements can be estimated from Equations (6.130), (6.180), and (6.187). The wave function $\Psi_2 = \exp[-W_0'/\hbar - W_1']$ is the semiclassical wave function localized in $\tilde{\mathbf{q}}_m$ and can be constructed in the same way as before. The principal exponent W_0', for instance, is explicitly given by

$$W_0'(\mathbf{q}) = \int_{\tau}^{\infty} \sum_i p_{oi}(\tau')\dot{q}_0^i(\tau')d\tau' \\ + \frac{1}{2}\sum_{ij} \tilde{A}_{ij}'(\tau)(q^i - q_0^i(\tau))\left(q^j - q_0^j(\tau)\right) + o\left((\Delta\mathbf{q})^2\right), \tag{7.37}$$

where $\tilde{A}'$ satisfies the same equation as Equation (6.97), but with the initial condition specified at $\tau = \infty$. Using the semiclassical form and neglecting the exponentially small contribution away from the instanton trajectory, we can estimate the matrix element as

$$I = N \exp[-\frac{S_0}{\hbar} - S_1] \int \sqrt{G} d\mathbf{q} f(\mathbf{q}) \exp[-\frac{\Delta\mathbf{q}\mathbf{A}\Delta\mathbf{q}}{\hbar}], \tag{7.38}$$

where $\mathbf{A} = (\tilde{\mathbf{A}} + \tilde{\mathbf{A}}')/2$ and $N = [\det A_m / G_m (\hbar\pi)^N]^{1/2}$ is the normalization factor of the semiclassical wave function. Equation (7.38) can be rewritten as a linear integral along the instanton path. Inserting the identity $1 = \int_{-\infty}^{\infty} d\tau\delta[\tau - \tau(\mathbf{q})]$ into the integrand and changing the order of integration, we obtain

$$I = N \exp\left[-\frac{S_0}{\hbar} - S_1\right] \int d\tau \sqrt{G\left(\mathbf{q}_0(\tau)\right)} f\left(\mathbf{q}_0(\tau)\right)$$
$$\times \int d\mathbf{s}\delta\left(\frac{\partial\tau}{\partial\mathbf{q}}\mathbf{s}\right) \exp\left(-\frac{\mathbf{s}^T \mathbf{A}[\mathbf{q}_0(\tau)]\mathbf{s}}{\hbar}\right) \tag{7.39}$$

up to the exponentially small terms. From Equation (6.179), we find $\partial\tau/\partial\mathbf{q} = \mathbf{p}_0/(\mathbf{p}_0^T \mathbf{g}\mathbf{p}_0)$, and the integration in Equation (7.39) gives

$$I = \sqrt{\frac{\det \mathbf{A}_m}{\hbar G_m \pi}} \exp\left[-\frac{S_0}{\hbar} - S_1\right] \int d\tau \sqrt{\frac{G}{\det \mathbf{A}}} \frac{(\mathbf{p}_0^T \mathbf{g}\mathbf{p}_0)}{\sqrt{\mathbf{p}_0^T \mathbf{A}^{-1} \mathbf{p}_0}} f(\mathbf{q}_0(\tau)). \tag{7.40}$$

Although the above derivations are rather straightforward, one cannot expect high accuracy from Equation (7.40). To see this, let us take $f(\mathbf{q}) = f_0 = \text{constant}$. In this case, the above estimation gives $I \propto f_0 \exp[-S_0/\hbar - S_1]$, while the exact result is definitely $I = 0$ as follows from the definition of Ψ_1 and Ψ_2. In other words, the loss of exact orthogonality introduces the error comparable to the matrix element itself. The estimation can be, however, improved by rewriting Equation (7.36) as

$$I = \langle\Psi_1|f - f(\mathbf{q}_r)|\Psi_2\rangle, \tag{7.41}$$

where $\mathbf{q}_r$ is an arbitrary reference point. This modification does not affect the exact value of the matrix element, but in the semiclassical approximation, it corresponds to the change $f[\mathbf{q}_0(\tau)] \to f[\mathbf{q}_0(\tau)] - f(\mathbf{q}_r)$ in the integrand of Equation (7.40).

The above method is applied to estimate the splitting of rotational constants in vinyl radical by using the results obtained at the *ab initio* CCSD(T)/aug-cc-pVTZ level. Diagonalization of the rotational metrics at the potential minimum gives the average value of three rotational constants and the corresponding eigenvectors $\mathbf{X}_k$ [see Equation (7.35)]. Then the splitting of rotational constant is estimated from Equations (7.31), (7.33), and (7.37). The results are presented in Table 7.7. In the present calculations, we took $\mathbf{q}_r = \mathbf{q}_b$, the symmetric midpoint of the instanton trajectory, which seems the only reasonable choice in the present problem. From Table 7.7 one can see that the semiclassical estimate reproduces the rotational constants and their splittings qualitatively, but absolute values are not very good compared to Δ_0, as expected.

TABLE 7.7
Rotational Constant (Average and Difference) in Vinyl Radical [in MHz]

method	A	B	C
CCSD(T)/aug-cc-pVTZ(average)	231800	33890	28500
(difference)	227	10.2	1.1
Experiment(average) Reference [149]	237065	32480	28438
(difference)	299	2.74	0.88

7.4 MALONALDEHYDE $C_3O_2H_4$

The tunneling effect in malonaldehyde (see Figure 6.2) has been well investigated both experimentally [155–160] and theoretically [41,42,161–170]. Malonaldehyde, as depicted in Figure 6.2, has an intramolecular hydrogen bond of O–H · · · O where the tunneling of a hydrogen atom between two oxygen atoms causes the splitting in the vibrational energy levels. The tunneling splitting has been determined experimentally by microwave spectroscopic studies. This is a heavy-light-heavy system and the reaction path is sharply curved near the transition state. Carrington and Miller [162] first estimated the ground-state splitting based on the reaction surface Hamiltonian formalism that utilized two large-amplitude coordinates, and further extensions to the three-dimensional model were made by Shida et al. [163,164]. Full-dimensional calculations have been carried out by the semiclassical trajectory method [165] and by the instanton theory [41,42]. A full-dimensional potential energy surface function was generated by Yagi et al. [166] by using the modified-Sheppard interpolation (MSI) technique [36,36,171–174,176,177] at the second Møller-Plesset perturbation theory [151] (MP2) level with 6-31G(d,p) basis set [152], which was then used within the semiclassical trajectory method of Makri and Miller [35] to estimate the tunneling splitting. Meyer and Ha [169] also constructed a full-dimensional potential energy surface by using the MP2/aug-cc-pVDZ method and estimated the tunneling splitting from the calculations of vibrational eigenenergies of the Hamiltonian along an approximate reaction path. Coutinho-Neto et al. carried out a full quantum mechanical calculations by using MSI potential energy function developed by Yagi et al. [166]. As was pointed out already, we have found out that the accuracy of the MP2/6-31G(d,p) potential energy surface is not good enough and the direct dynamical *ab initio* method is the only way to get a good agreement with experiment. Besides, the instanton trajectory can be determined accurately in any high-dimensional system with use of a high level of quantum chemical *ab initio* method.

Before presenting the most accurate treatment based on the high-level *ab initio* quantum chemical calculations, we first investigate the accuracy of the analytical full-dimensional potential energy surface given by Sewell et al. [165]. In planar configuration this potential energy surface differs from the one reported previously [178]. It was found that it is necessary to modify the planar part given in Reference

[165] according to the older potential energy surface from Reference [178] in order to reproduce the normal mode frequencies. Namely, the implemented potential energy surface has the form

$$S_1(R)V_A + [1 - S_1(R)]V_B, \tag{7.42}$$

where R is the reaction coordinate $r_{\mathrm{OH}} - r_{\mathrm{O\cdots H}}$, and V_A and V_B are the zeroth-order potentials of two symmetric isomers. The parameter of the switching function $S_1(R)$ (Equation (2) of Reference [165]) was taken from Reference [178] with $a = 6.0\,\mathring{A}^{-1}$. The Morse part of the zeroth-order potential (Equation (3) of Reference [165]) was also modified according to Reference [178]. The second switching function $S_2(R)$ (Equation (4) of Reference [165]) was omitted from the Morse functions except for the two bonds r_{OH} and $r_{\mathrm{O\cdots H}}$.

As was explained in Sections 6.2 and 6.3, the instanton trajectory was determined by using the 21 independent Cartesian coordinates in the body-fixed frame. The Hamiltonian is given in the same form as Equation (6.128). The initial guess of the instanton trajectory was taken as a straight line connecting two potential minima, and the converged result was obtained after 32 iterations in space of 20 basis functions. The calculations were repeated with 30 basis functions, but the result was identical. The final values obtained are $\Delta_0(\mathrm{H}) = 57.7$ cm^{-1} for the hydrogen tunneling, and $\Delta_0(\mathrm{D}) = 8.63$ cm^{-1} for the deuterium substitution [43]. These values are different from the corresponding values, $\Delta_0(\mathrm{H}) = 21.8$ cm^{-1} and $\Delta_0(\mathrm{D}) = 5.2$ cm^{-1}, reported in Reference [178]. We believe that the disagreement between the present results and the experimental values [$\Delta_0(\mathrm{H}) = 21.6$ cm^{-1} and $\Delta_0(\mathrm{D}) = 2.915$ cm^{-1}] [157,159] indicates the insufficiency of the potential energy surface function. It is interesting to note that the isotope effect, $\Delta_0(\mathrm{H})/\Delta_0(\mathrm{D})$, of the present result ($\simeq 6.7$) is closer to the observed value ($\simeq 7.4$).

Now, we apply several *ab initio* quantum chemical methods [122,123]. The employed methods are the MP2, QCISD [153], and CCSD(T) [179], with the basis sets of Dunning's cc-pVDZ [124] and aug-cc-pVDZ [124,180]. We also attempted the hybrid basis set of aug-cc-pVDZ (for two oxygen atoms and transferring hydrogen atom, H_9 in Figure 6.2) and cc-pVDZ (for other atoms), denoted as (aug-)cc-pVDZ. *Ab initio* calculations were carried out by using GAUSSIAN-98 [181] and MOLPRO [182]. Table 7.8 gives a summary of barrier height obtained by each method as well as that by using the MSI [MP2-6-31G(d,p)] potential energy surface [166]. Comparisons of the results obtained by cc-pVDZ with those by aug-cc-pVDZ show that calculations with augmented basis sets yield larger barriers by 0.38 and 0.61 kcal/mol at the MP2 and QCISD levels, respectively. The results with aug-cc-pVDZ and (aug-)cc-pVDZ are almost the same at both MP2 and QCISD levels, which indicates that the augmented basis sets are necessary to be added only on two oxygen atoms and the transferring hydrogen atom. Then, the coupled-cluster singles and doubles including a perturbational estimate of triple excitations [CCSD(T)] method with the (aug-)cc-pVDZ basis sets and with the hybrid basis set of aug-cc-pVTZ and cc-pVTZ, denoted as (aug-)cc-pVTZ, is applied to locate the stationary points. Table 7.8 shows the resulting barrier height at the CCSD(T)/(aug-)cc-pVDZ level as 4.53 kcal/mol, which is in between those at MP2 and QCISD obtained as 3.31 and 5.44 kcal/mol, respectively. The best estimation of the barrier height was obtained as

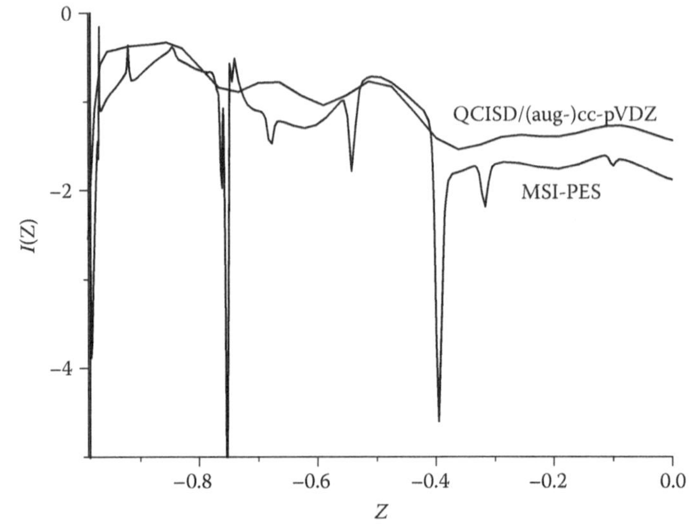

FIGURE 7.8 The integrand $I(z)$ for the second pre-exponential factor S_1 in the case of malonaldehyde. The curve with sharp peaks corresponds to the interpolated global potential energy surface from Reference [166]. The smooth curve is obtained without global interpolation of the *ab initio* data. (Taken from Reference [122] with permission.)

3.81 kcal/mol at the CCSD(T)/(aug-)cc-pVTZ level of theory; the effect of triple-ζ level of basis sets decreases the barrier height. Computations at the level of CCSD(T)/(aug-)cc-pVTZ are, however, very much time consuming and in the following calculations the best level of theory employed is CCSD(T)/(aug-)cc-pVDZ.

We first computed Δ_0 for MSI potential energy surface function reported recently [166]. In this case the potential is given in the analytical form and the gradient and Hessian can be easily evaluated numerically. The whole numerical procedure is the same as that mentioned above [43]. We used 30 basis functions to describe the instanton trajectory and certified five stable digits in the action. The present theory gives $\Delta_0(\mathrm{H}) = 30.7\ \mathrm{cm}^{-1}$ for H transfer and $\Delta_0(\mathrm{D}) = 4.58\ \mathrm{cm}^{-1}$ for the deuterium substitution, which deviate quite a bit from the experimental values mentioned above. Insufficient accuracy of the quantum chemical method, i.e., MP2/6-31G(d,p), is found to be responsible for this disagreement. Another crucial cause of the error is incorrect topology of the potential energy surface in the vicinity of the instanton. This can be seen from Figure 7.8, which shows the behavior of the integrand

$$I(z) = \frac{1}{z}\left(\mathrm{Tr}[\mathbf{A}(z) - \mathbf{A}_m]\right) \tag{7.43}$$

for the extra exponential terms $S_1 = 2W_{1\Sigma}$ [see Equation (6.100)]. In Figure 7.8 we compare $I(z)$ for the global potential energy surface and for the direct *ab initio* method at the level of QCISD. The sharp peaks in Figure 7.8 indicate erroneous behavior of the local frequencies and the obtained semiclassical solution along the instanton path. This is caused by the interpolation of the *ab initio* reference points away from the

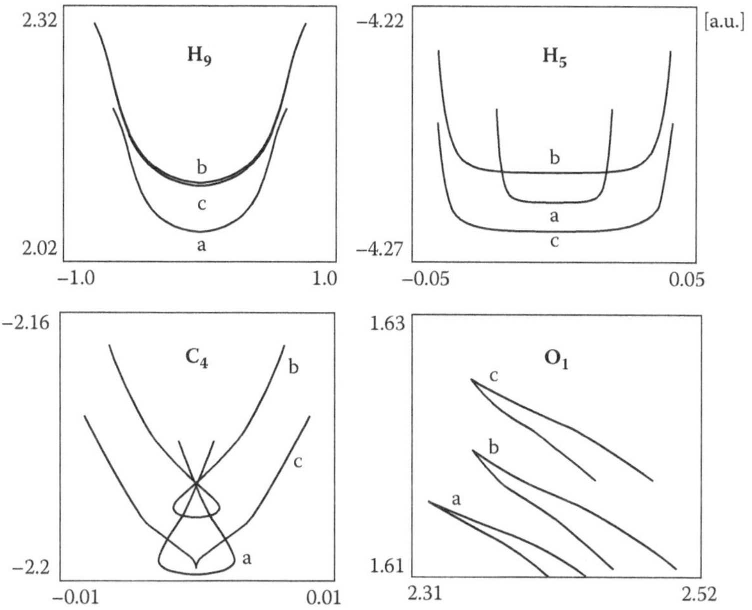

FIGURE 7.9 Instanton trajectory in the malonaldehyde obtained by the direct *ab initio* calculations without global interpolation: (a) MP2/cc-pVDZ, (b) QCISD/(aug-)cc-pVDZ, and (c) CCSD(T)/(aug-)cc-pVDZ. (Taken from Reference [122] with permission.)

instanton. This inherent drawback of the global interpolation is remedied by the method explained in Chapter 6. Namely, we reconstruct $V(\mathbf{Q}_0)$, $(\partial V/\partial q^j)[\mathbf{q}_0(z)]$ and $(\partial^2 V/\partial q^i \partial q^j)[\mathbf{q}_0(z)]$ as functions of the one-dimensional parameter z and calculate the action and its derivatives [see Equation (6.115)].

We computed the tunneling splitting for the MP2/cc-pVDZ, QCISD/(aug-)cc-pVDZ, and CCSD(T)/(aug-)cc-pVDZ methods. The computation of the energy, gradient, and Hessian at one point took 14 minutes, 1 day 3 hours 43 minutes, and 10 days 16 hours 28 minutes of CPU time on a Pentium 4.18-GHz Linux PC computer for MP2/cc-pVDZ, QCISD/(aug-)cc-pVDZ, and CCSD(T)/(aug-)cc-pVDZ, respectively. We used the instanton trajectory in the case of MSI potential energy surface as an initial guess for MP2/cc-pVDZ calculations, and the converged $\mathbf{q}_0(z)$ is used as the first approximation for the QCISD/(aug-)cc-pVDZ level, etc. In all these cases, four extra iterations were good enough to achieve convergence at each higher *ab initio* level. Figures 7.9 and 7.10 depict the instanton trajectory. For the purpose of illustration, the motion of four different atoms in the molecular plane is shown. The instanton path is naturally a smooth line in the 21-dimensional $\mathbf{q}$ space and thus the sharp curvature and self-crossings in Figures 7.9 and 7.10 are due to the projection onto the two-dimensional plane. Figure 7.9 compares the paths for the above-mentioned three different *ab initio* methods. In Figure 7.10 we show the instantons for the global MP2 level [166] and the semiempirical analytical potential energy surface [165] in comparison with the accurate *ab initio* potential. The shape of the path in the semiempirical

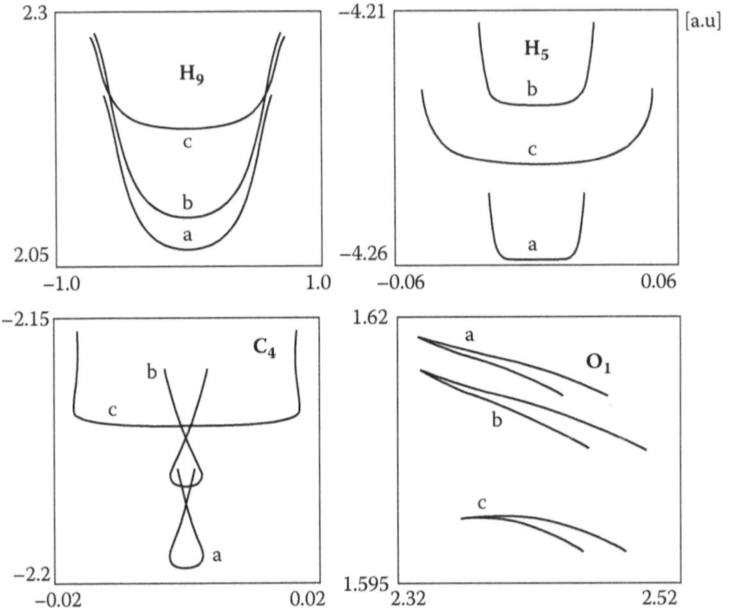

FIGURE 7.10 Comparison of the *ab initio* instanton trajectory [the same as (c) in Figure 7.9] with the one obtained with the global MP2 potential (b) and semiempirical analytical potential (c) in the case of malonaldehyde. (Taken from Reference [122] with permission.)

case drastically differs from the correct one. This is clearly seen also for the large amplitude motion of the tunneling hydrogen atom H_9. The reasonable value of the tunneling splitting obtained [43] is due to mutual cancellation of the short tunneling path and the overestimated barrier height. As shown in Table 7.8, the best values of the tunneling splitting obtained, $\Delta_0(H) = 21.2 - 22.2\ cm^{-1}$ and $\Delta_0(D) = 3.0\ cm^{-1}$, are in very good agreement with the experimental values, $\Delta_0(H) = 21.6\ cm^{-1}$ and $\Delta_0(D) = 2.9\ cm^{-1}$. The MP2/cc-pVDZ and QCISD/(aug-)cc-pVDZ methods are not accurate enough and both give about four times discrepancy compared with the experimental value. This deviation is mainly due to the incorrect barrier height, which in MP2/cc-pVDZ and QCISD/(aug-)cc-pVDZ methods constitutes 2.91 and 5.41 kcal/mol, respectively. The correct value lies in between these two. The error also correlates with the behavior of the potential along the instanton path as shown in Figure 6.1. However, the deviation of the results for the different *ab initio* methods mostly comes from the principal exponential factor S_0, which strongly depends on the barrier height. The pre-exponential factors are not that sensitive, which indicates that the different *ab initio* methods produce potential functions with similar *topology*. As a result, the *shapes* of the instanton trajectories are close to each other. The main differences among the three trajectories in Figure 7.9 are actually due to their starting positions (potential minima). This explains the fact that at the higher *ab initio* computational level convergence is easily achieved with a few extra steps of iteration. Thus, although the MP2/cc-pVDZ and QCISD/(aug-)cc-pVDZ calculations cannot

TABLE 7.8
Barrier Height (BH) for the Proton Transfer Reaction in Malonaldehyde Computed at the MP2, QCISD, and CCSD(T) Level of Quantum Chemistry with Various Types of Basis Sets. Three of These Methods Were Used for Direct Determination of the Instanton Trajectory. The Resulting Tunneling Splitting Δ_0 as Well as the Values of the Three Constituents B', S_0, and S_1 in the Expression $\Delta_0 = B' \exp[-S_0/\hbar - S_1]$. The Splitting for the Deuterium Substitution Is Shown for MSI and Mp2/cc-pVDZ. The Isotope Effect Is Also Estimated by Using the Highest Level of the CCSD(T)/(aug-)cc-pVTZ Method for S_0 and by Taking B' and S_1 from MP2/cc-pVDZ and CCSD(T)/(aug-)cc-pVDZ

Method/basis set	Δ_0(cm^{-1})	BH (kcal/mol)	B'(cm^{-1})	S_0	S_1
MP2/6-31G(d, p)(MSI)[a]	30.7	3.62	1855	5.56	−1.46
	4.58(D)		1408	7.09	−1.31
MP2/cc-pVDZ	77	2.91	1642	4.53	−1.48
	14.9(D)		1269	5.74	−1.30
/aug-cc-pVDZ		3.29			
/(aug-)cc-pVDZ[b]		3.31			
QCISD/(aug-)cc-pVDZ	4.5	5.44	2200	7.38	−1.18
/cc-pVDZ		4.80			
/aug-cc-pVDZ		5.41			
CCSD(T)/(aug-)cc-pVDZ	16.4	4.53	1922	6.14	−1.37
/(aug-)cc-pVTZ[c]	21.2[d]	3.81	1642	5.83	−1.48
	22.2[d]		1922	5.83	−1.37
	3.0(D)		1269	7.35	−1.30
Experiment	21.6[e]				
	2.9(D)[f]				
Tautermann CCSD(T)	24.7[g]				
Empirical	57.7[h]				
	8.63(D)[h]				

[a] Reference [166].
[b] aug-cc-pVDZ for two oxygen atoms and for the transferring hydrogen atom, and cc-pVDZ for the other atoms.
[c] aug-cc-pVDZ for two oxygen atoms and for the transferring hydrogen atom, and cc-pVTZ for the other atoms.
[d] Δ_0 is estimated by taking B', S_1 from MP2/cc-pVDZ(4,6) and CCSD(T)/(aug-)cc-pVDZ(5).
[e] Reference [159].
[f] Reference [156].
[g] Reference [42].
[h] Reference [41].

produce a reliable value of tunneling splitting, they enable us to reduce the numerical efforts on the higher *ab initio* levels. In the present case, the instanton trajectory for CCSD(T)/(aug-)cc-pVDZ was obtained by four extra steps of iteration by using only 33 *ab initio* points in total. We have obtained the tunneling splitting 16.4 cm^{-1} in a relatively good agreement with the experiment. As shown in Table 7.8, the potential barrier is overestimated in the CCSD(T)/(aug-)cc-pVDZ method, which is likely to be the reason for the conservative calculated value. To investigate the accuracy of the above result, we estimated the splitting by using the higher CCSD(T)/(aug-)cc-pVTZ level of electronic structure theory. The full implementation with this *ab initio* level is still too time consuming. As noted above, however, instanton paths for different *ab initio* methods show fairly similar topology; thus we can estimate Δ_0 by using the

CCSD(T)/(aug-)cc-pVTZ energy along the previously obtained instanton path. This gives an action S_0 at the higher *ab initio* level while the other two factors B' and S_1 can be taken from the lower level. In this way, we have obtained $\Delta(\mathrm{H}) = 21.2(22.2)\ \mathrm{cm}^{-1}$ by taking the pre-exponential factors from the MP2 (CCSD) calculations. The isotope effect was estimated in the same way. For the deuterium substitution of malonaldehyde we first found the accurate instanton trajectory and calculated the splitting by the MP2/cc-pVDZ *ab initio* method. The action S_0 was recalculated for this path by using CCSD(T)/(aug-)cc-pVTZ *ab initio* points while the other two factors B' and S_1 remained unchanged. This gives $\Delta_0(\mathrm{D}) = 3.0\ \mathrm{cm}^{-1}$, as mentioned above, in very good agreement with the experimental data. In the present work, we first resorted to a hybrid scheme of the calculations by using the MSI potential obtained by MP2/6-31G. Due to the absence of such a potential energy surface, this preliminary step seems impossible in general. But actually, this step is not needed. By simply taking the straight line connecting two potential minima as the initial guess and using any non-time-consuming *ab initio* method, as was done before, it is possible to accomplish this preliminary step easily after 10–20 iterations with 100–120 reference points in total. Thus the obtained path can now be taken as the initial approximation for the highest possible *ab initio* level. This second stage usually requires only three to four extra steps of iteration to obtain the final results.

Finally, we demonstrate a strong effect of out-of-plane vibrational motions on the tunneling splitting [183]. In order to do this, first we have to define 15 internal coordinates to describe in-plane motions and construct the corresponding Hamiltonian. Following the same method as that explained in Section 6.3.1, we introduce Cartesian coordinates x_{in} in the body-fixed (BF) frame of reference,

$$\mathbf{r}_n = x_{1n}\mathbf{e}_1 + x_{2n}\mathbf{e}_2 \ (n = 1, 2, \ldots, 9), \tag{7.44}$$

where n numerates the atoms (see Figure 6.2). The origin of the BF frame is set to the center of mass,

$$\sum_{n=1}^{9} M_n x_{in} = 0 \ \ (i = 1, 2), \tag{7.45}$$

where M_n is the mass of the nth atom. The orientation of the BF axis $(\mathbf{e}_1, \mathbf{e}_2)$ is determined by taking $\mathbf{e}_1$ in the direction parallel to a vector connecting O_1 and O_8. In terms of x_{in} this gives

$$x_{21} = x_{28}. \tag{7.46}$$

The three constraints given by Equations (7.45) and (7.46) define the BF system. The independent internal coordinates $q^k (k = 1, 2, \ldots, 15)$ are chosen to be 14 Cartesian coordinates of seven atoms, O_1, C_2, H_3, C_4, C_6, H_7, and H_9, and x_{18} (x coordinate of O_8). The kinetic metric defined as

$$ds^2 = \sum_n M_n d\mathbf{r}_n^2 \tag{7.47}$$

is rewritten by performing a transformation $\{d\mathbf{r}_n\} \rightarrow \{dq^k, d\omega, d\mathbf{R}_{cm}\}$, where $d\boldsymbol{\omega} = \mathbf{e}_3 d\omega$ represents an infinitesimal rotation in the plane and $\mathbf{R}_{cm}$ is the center-of-mass position that separates from the equation of motion and can be put to zero. Inserting

$$d\mathbf{r}_n = \sum_k \frac{\partial \mathbf{r}_n}{\partial q^k} dq^k + [d\omega \times \mathbf{r}_n] \tag{7.48}$$

into Equation (7.47), we obtain

$$ds^2 = \sum M_n \left(\frac{\partial x_{in}}{\partial q^k} \frac{\partial x_{in}}{\partial q^l} dq^k dq^l + 2 \left[x_{in} \frac{\partial x_{2n}}{\partial q^l} - x_{2n} \frac{\partial x_{1n}}{\partial q^l} \right] dq^l d\omega \right.$$
$$\left. + \left[\mathbf{r}_n^2 \delta_{ij} - x_{in} x_{jn} \right] d\omega^2 \right). \tag{7.49}$$

Equation (7.49) determines the covariant metric tensor $G_{ij}(\mathbf{q})$ in terms of 16 generalized coordinates $\{d\mathbf{q}^T, d\omega^T\}$ and we have

$$ds^2 = (\mathbf{q}^T, d\omega^T) \begin{pmatrix} G_{qq} & G_{q\omega} \\ G_{\omega q} & G_{\omega\omega} \end{pmatrix} \begin{pmatrix} d\mathbf{q} \\ d\omega \end{pmatrix}. \tag{7.50}$$

The kinetic energy operator T for the total angular momentum $J = 0$ is given by

$$T = \frac{\hbar^2}{2\sqrt{G}} \sum_{i,j=1}^{15} \frac{\partial}{\partial q^i} \left(\sqrt{G} g^{ij} \frac{\partial}{\partial q^i} \right), \tag{7.51}$$

where g^{ij} is the 15×15 block of the inverse G^{ij} of the covariant tensor which satisfies

$$\sum_{k=1}^{16} G_{ik} G^{kj} = \delta_i^j \tag{7.52}$$

and G is the determinant of the metric. The tunneling splitting Δ_0 is evaluated as before by the general formula [see Equations (6.100) and (6.117), where $B = B' \exp[-S_1]$, $S_0 = 2W_{0\Sigma}$, and $S_1 = 2W_{1\Sigma}$],

$$\Delta_0 = B' \exp\left[-\frac{S_0}{\hbar} - S_1 \right]. \tag{7.53}$$

Here S_0 is the action along the instanton trajectory and the two additional factors, B' and S_1, are given by

$$B' = \sqrt{\frac{4G_\Sigma \det \mathbf{A}_m}{\pi G_m \det \mathbf{A}_\Sigma} \frac{(\mathbf{p}_0^T \mathbf{g} \mathbf{p}_0)_\Sigma}{\sqrt{(\mathbf{p}_0^T \mathbf{A}^{-1} \mathbf{p}_0)_\Sigma}}} \tag{7.54}$$

and

$$S_1 = \int_{-\infty}^{0} d\tau [\mathrm{Tr}(\mathbf{A} - \mathbf{A}_m)]. \tag{7.55}$$

The matrix $\mathbf{A}$ is the same as that defined before in Chapter 6.

TABLE 7.9
Tunneling Splitting of the Ground State (in cm^{-1}) in the Planar Model of Malonaldehyde with Use of Various *ab initio* Methods

Method	Planar model (15D)	Full D model (21D)
MSI-pot.func.	71.2	30.7[a]
MP2/cc-pVDZ	171	77[a]
QCISD/(aug-)cc-pVDZ	8.1	4.5[a]
CCSD(T)/(aug-)cc-pVDZ	160	16.4[a]
Experiment		21.6[b]

[a]Reference [122].
[b]Reference [160].

For the calculations of tunneling splitting, two types of potential energy surfaces are used. One is the full-dimensional potential energy function MSI developed by Yagi et al. [166] based on the MP2/6-31G(d,p) level of theory. The other is the *ab initio* data calculated at every point required in the present treatment explained in Section 6.3. The *ab initio* methods used are MP2/cc-pVDZ, QCISD/(aug-)cc-pVDZ, and CCSD(T)/(aug-)cc-pVDZ. The augmented basis sets were used only for the OH $\cdots$ O fragment. The detailed scheme is the same as that explained before. Table 7.9 presents the ground-state tunneling splitting for the planar malonaldehyde for various potential energy surfaces. One can see that the planar model gives much larger tunneling splitting than the full-dimensional one. It should be noted that the result of the CCSD(T) level, which is of the best quality, gives nearly 10 times larger splitting. With use of MSI surface, the planar model gives larger splitting by a factor of two than the full-dimensional one. These results indicate that the out-of-plane degrees of freedom play an important role in tunneling dynamics. In order to understand the main origin of this deviation, in Table 7.10 we show the values of the three parameters B', S_0, and S_1 in Equation (7.53) for the planar model and full-dimensional model. Note that the action S_0 is the same for the two models, since the instanton path is the same for the two cases. It is seen that both B' and S_1 contribute to a larger splitting in the planar model. The value of S_1 is largely decreased with use of CCSD(T) potential energy surface, which is actually the main reason that the planar model with CCSD(T) gives very large splitting.

As explained before, the quantity S_1 is calculated with use of the parameter z [see Equation (7.9)] as

$$S_1 = \int_{-1}^{0} I(z)dz \equiv \int_{-1}^{0} \frac{dz}{\dot{z}} \left[\mathrm{Tr}(\mathbf{A} - \mathbf{A}_m) \right]. \tag{7.56}$$

Figure 7.11 illustrates the behavior of the integrand $I(z)$ in this equation obtained by the CCSD(T) method. The lower and upper curves correspond to the planar and

TABLE 7.10
The Values of B', S_0, and S_1 in Equation (7.53) for the Planar Model of Malonaldehyde Calculated at Various *ab initio* Level of Theory

Method	B'	S_0[a]	S_1
MP2/cc-pVDZ			
Planar model	2098	4.53	−2.03
Full-dimensional model[b]	1642	4.53	−1.48
QCISD/(aug-)cc-pVDZ			
Planar model	3093	7.38	−1.43
Full-dimensional model[b]	2200	7.38	−1.18
CCSD(T)/(aug-)cc-pVDZ			
Planar model	2843	6.14	−3.26
Full-dimensional model[b]	1922	6.14	−1.37

[a]Note that S_0 is the same for the two models, since the instanton path is the same.
[b]The data of full-dimensional model are taken from Reference [122].

the full-dimensional configuration, respectively. One can see from this figure that the integrand for these two cases is very different in the vicinity of $z = -1$. The value at $z \simeq -1$ is completely determined by the anharmonicity of the potential in the vicinity of the potential minimum. Figure 7.11 suggests that the anharmonic coupling

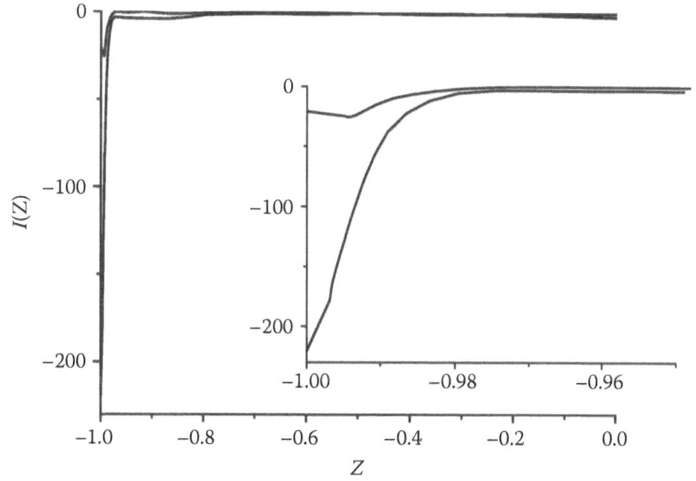

FIGURE 7.11 Behavior of the integrand $I(z)$ in Equation (7.56). The lower and upper curves correspond to those of the planar and full-dimensional models of malonaldehyde, respectively. (Taken from Reference [183] with permission.)

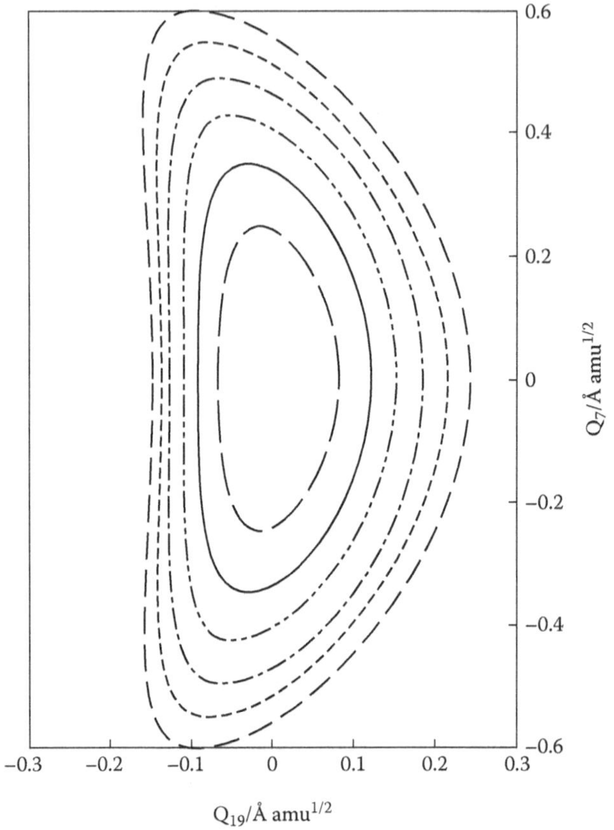

FIGURE 7.12 Contour plot of the potential energy surface with respect to 7th and 19th normal coordinates (Q_7 and Q_{19}) of malonaldehyde calculated at the level of MP2/cc-pVDZ. Q_7 and Q_{19} represent the out-of-plane vibration of the tunneling hydrogen atom and the OH stretching vibration, respectively. The contour increment is set to 800 cm^{-1}. See also Figure 3 of Reference [168] and discussion therein. (Taken from Reference [183] with permission.)

between the out-of-plane and in-plane modes is stronger and the energy exchange is much faster near the equilibrium structure.

Figure 7.12 gives a contour plot of the potential energy surface with respect to 7th and 19th normal coordinates at the equilibrium structure (Q_7 and Q_{19}, respectively) obtained at the level of MP2/cc-pVDZ. The point (Q_7, Q_{19}) = (0.0, 0.0) is the global minimum. The calculation has been carried out by using GAUSSIAN-98 program package [181]. Q_7 represents the out-of-plane vibration of the tunneling hydrogen atom and Q_{19} represents the OH stretching vibration. The positive direction of Q_{19} is taken such that the OH bond is elongated. We note that the energy profile of this section is also discussed in [168]. In Figure 7.12 a strong anharmonic coupling between the two modes is observed. This feature of the potential energy surface indicates that the OH stretching vibrational energy flows to the out-of-plane vibration when the OH

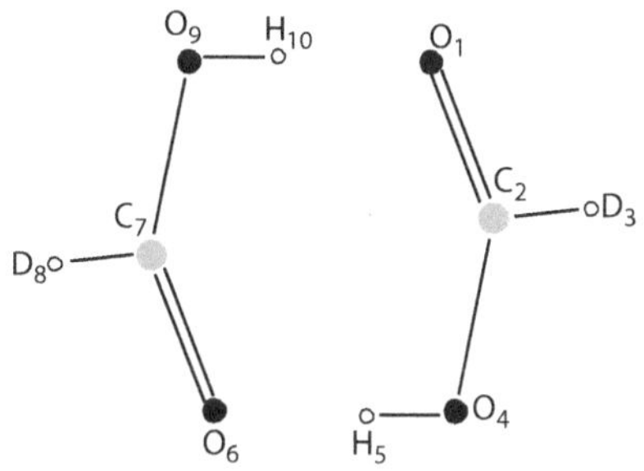

FIGURE 7.13 Geometry of the formic acid dimer (FAD) with labeling as used in the text. (Taken from Reference [184] with permission.)

bond is contracted. In the planar model without such couplings, the large energy stays in the OH stretching mode and thus the rate of tunneling increases. In the heavy-light-heavy system like in the hydrogen transfer, the reaction path is sharply curved in general because of the mass disparity, which induces a strong coupling between the reaction coordinate and the transverse vibrational coordinates. Therefore, in such a situation, the often-used adiabatic description would be inadequate. In the one-dimensional model, the transversal part in the pre-exponential factor is completely absent, which can even change the order of magnitude of Δ_0 in polyatomic molecules. The above results demonstrate that proper treatment of multidimensional effects is very important even for the qualitative level of estimation.

7.5 FORMIC ACID DIMER $(DCOOH)_2$

Carboxylic acid dimers take the role of malonaldehyde as a standard system, when it comes to symmetric *double* hydrogen bonds. There has been considerable work devoted to this system ranging from the investigation of the infrared spectra [185–188] to NMR (nuclear magnetic nesonance) relaxation [189] and neutron scattering [190] experiments that established the importance of tunneling in these systems. The formic acid dimer (FAD) shown in Figure 7.13, as the simplest carboxylic acid dimer, has been particularly attractive for theoretical studies. Various quantum chemical methods have been applied to predict geometries, infrared and Raman spectra, dimerization energies, as well as double proton-transfer barriers (see, for instance, References [191–193]).

The double hydrogen bond in principle facilitates two mechanisms of double proton transfer: stepwise and concerted. Studying the transfer rates within the semiclassical tunneling approximation of the variational transition state theory, Kim found that the two protons are transferred synchronously across the transition state with D_{2h} symmetry [194]. The actual proton-tunneling distance is considerably reduced due to the contraction of the hydrogen bonds, i.e., heavy atom motion promotes the proton transfer. On the other hand, the *ab initio* path-integral Car-Parinello calculations predict that the motion of the two protons in the vicinity of the potential minima is asynchronous, whereas it becomes synchronous at the transition state. Furthermore, these studies have confirmed that the quantum fluctuations may lead to considerable deviation from the minimum energy path [195]. A similar conclusion is reached from the *ab initio* classical molecular dynamics study by Ushiyama and Takatsuka [196]. They have observed that there is a certain time lag of about 8 fs between the two proton transfer events. In contrast, Wolf et al. concluded from their classical molecular dynamics study that the double proton transfer is mostly synchronous [197]. The first calculation of the ground-state tunneling splitting, which takes into account the multidimensionality on the basis of the reaction surface Hamiltonian method, was done by Shida et al. [198]. They obtained the value of $\Delta_0 = 0.004$ cm^{-1}. More recently, Tautermann et al. applied their approximate version of instanton theory, which is based on the earlier work by Garg [199], to the tunneling splittings for various carboxylic acid dimers; for FAD they obtained $\Delta_0 = 0.0022$ cm^{-1} [200]. These values and the underlying theoretical approaches can be tested against the recent experiment done by Madeja and Havenith [201], in which the high-resolution vibration-rotation spectra in the region of the C-O stretching vibration around 1245 cm^{-1} has been reported. From their data, the ground-state tunneling splitting is extracted as $\Delta_0 = 0.00286$ cm^{-1}. Furthermore, it was found out that upon the C-O stretching excitation the tunneling splitting increases by a factor of 3.49 to 0.00999 cm^{-1}. The acceleration of tunneling upon vibrational excitation is quite natural and has been observed in many systems as reported before. However, the splitting can actually decrease or oscillate against energy. In the case of FAD, Smedarchina and co-workers argued that the C-O vibration is not of the usual promoting type [202]. Undoubtedly, vibrationally assisted tunneling is an important issue and, in fact, the interpretation of experimental data should rely on the correct assignment of ground- and excited-state tunneling splitting. Using their version of instanton theory, Smedarchina et al. concluded that the original experimental assignment has to be reversed. Their calculated ground-state splitting is 0.0147 cm^{-1}, which is only slightly higher than the proposed new experimental assignment that gives 0.0125 cm^{-1} [201]. As is discussed below, however, the original assignment by Madeja and Havenith [201] seems correct.

Hereafter, the numerical calculations and their results on the deuterated formic acid dimer $(DCOOH)_2$ (see Figure 7.13) are presented for both vibrational ground and excited states with use of the theory provided in Chapter 6 [203]. To construct the rotation-free ($J = 0$) quantum Hamiltonian for FAD we follow the procedure in Section 6.3.1. All atoms are enumerated as shown in Figure 7.13 and we introduce thirty Cartesian coordinates $\mathbf{r}_n = \sum_{i=1}^{3} x_{in}\mathbf{e}_i (n = 1, 2, \ldots, 10)$. In the body-fixed frame of reference, six conditions should be imposed. The first three conditions put the origin in the center of mass. In the present calculations we choose the body-fixed axes in such a way that the atom H_{10} lies in X-Y plane and the body-fixed axis is oriented along the $O_1 - O_9$ line. Using these six constraints we can take any 24 coordinates x_{in} as independent internal coordinates and construct the metric tensor in the same as before. Note that in the case of double proton transfer, such choice of internal coordinates does not guarantee the symmetry of instanton trajectory with respect to the exchange of H_5 and H_{10}, which provides an additional accuracy check of the calculations.

All results of the tunneling splittings for various levels of quantum chemical calculations are summarized in Table 7.11, where we also give the values of the parameters S_0, B', and S_1 [see Equation (7.53)]. For the low level of quantum chemical calculations we have used the density-functional theory (DFT) with the B3LYP exchange-correlation functional. All calculations are done using the GAUSSIAN03 suite of programs [154]. Within the use of small basis set such as 6-31+G(d), this method has been shown to provide quite accurate barrier height for FAD (2933 cm^{-1}) as compared with the high-level methods (see also References [200,204]). The instanton trajectory was first calculated at this level of potential energy surface by using the method of Section 6.3.1. Convergence was obtained after nine iteration steps with each one consisting of 12 reference points. We have confirmed four stable significant digits in the final value of the classical action S_0. Figure 7.14 shows the projection of the instanton path onto the X-Y plane for the two tunneling hydrogen atoms and for $O_1 - O_9$ distance. It should be noted that the symmetry of this picture confirms the accuracy of the present calculation. The ground-state tunneling splitting for this instanton trajectory is 0.0077 cm^{-1}, that is, about three times larger than the experimental value (see Table 7.11). Using the coupled cluster approach with approximate triple excitations [CCSD(T)], we have introduced a correction to the potential energy surface. As can be seen from Table 7.11, however, the agreement with experiment gets worse. This merely indicates that the basis set is too small to give any reasonable result at the CCSD(T) level. In fact, the barrier becomes as high as 4515 cm^{-1}, which gives rather small tunneling splitting. This instanton trajectory is further used as a trial path for the variational calculations using more flexible basis sets. In particular, we have determined the instanton trajectory at the DFT/B3LYP level with 6-311++G(3df,3pd) as well as an augmented correlation-consistent polarized valence triple-zeta (aug-cc-pVTZ) basis set. In both cases the convergent trajectory was obtained after three extra steps of iteration. The barrier heights are similar to each other, that is, 2280 and 2251 cm^{-1}. However, the DFT tunneling splittings are about two orders of magnitude too large in both cases (0.17 and 0.23 cm^{-1}). As is seen in Figure 7.14, the two hydrogen atoms approach to each other more closely compared to the case of small basis set. For the two different large basis sets there appears to be no difference (only small offset) in the motion of hydrogens (oxygens). At the CCSD(T) level, geometry optimization for the 10 atomic FAD and subsequent iterative calculations of Hessian is still too time consuming. Therefore, we have estimated Δ_0 by using the energies from the higher level with geometry of instanton path at the lower level. In doing so, we have neglected the difference in the equilibrium configurations of the DFT and CCSD(T) potential energy surfaces. This CCSD(T) correction results in barrier heights of 2837 and 2771 cm^{-1} for the two basis sets and the corresponding tunneling splittings are 0.0038 and 0.0044 cm^{-1}, which are in relatively good agreement with the experimental value of 0.0029 cm^{-1} [201]. This result indicates that the original assignment by Madeja and Havenith [201] is correct. In passing, we note that the combination of the DFT/6-31++G(3df,3pd) instanton with the CCSD(T)/aug-cc-pVTZ energy corrections gives a splitting of 0.0042 cm^{-1}.

Unfortunately, the reliable final accurate value of Δ_0 cannot be obtained, since the highest possible quantum chemical calculations are not yet satisfactory enough

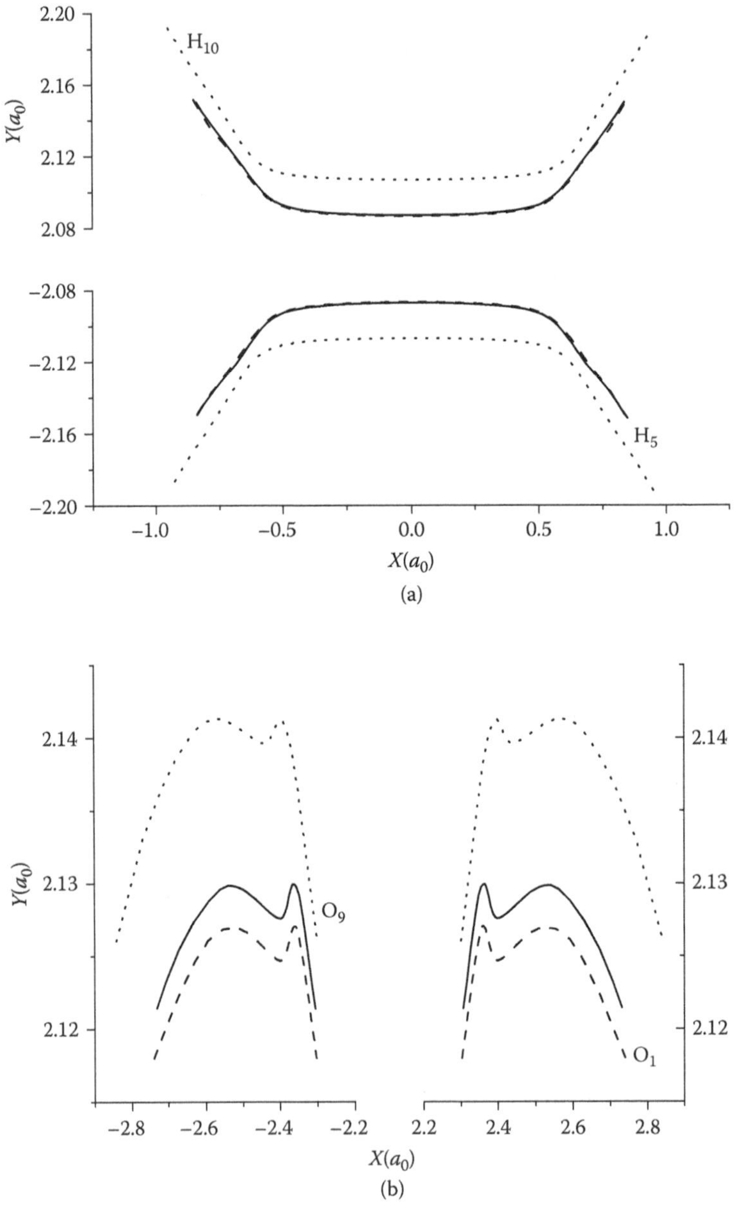

FIGURE 7.14 Instanton trajectory in FAD for three different *ab initio* methods [Dotted line: DFT/6-31+G(d), Solid line: DFT/aug-cc-pVTZ, Dashed line: DFT/6-311++G(3df,3pd)]. (a) Projection onto (X, Y) plane for hydrogen atoms H_5 and H_{10}. (b) The same for O_1 and O_9 (for the labeling see Figure 7.12). (Taken from Reference [184] with permission.)

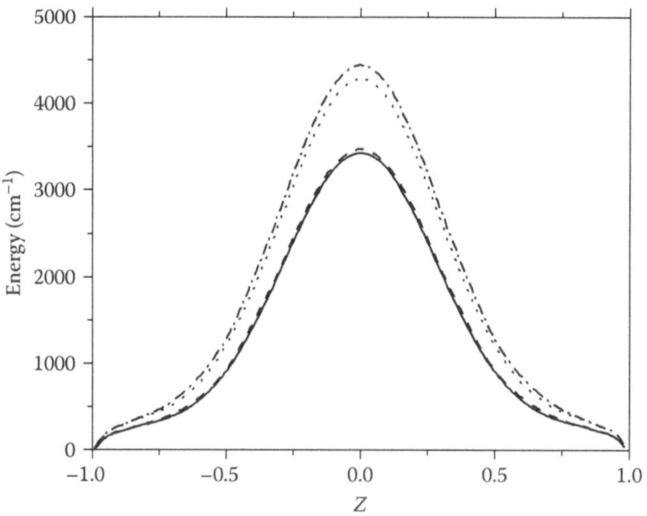

FIGURE 7.15 Potential energy along the instanton trajectory based on interpolation of a small number of *ab initio* points in the case of formic acid dimer. Dotted line: DFT/6-31+G(d); Solid line: DFT/aug-cc-pVTZ; Dashed line: DFT/6-311++G(3df,3pd); Dash-dotted line: CCSD(T) correction to DFT/aug-cc-pVTZ instanton. (Taken from Reference [184] with permission.)

because of the formidable CPU time. However, all the pre-exponential factors in Table 7.11 are close to each other, which confirms the premise about the similarity of the *shapes* of instanton paths. It is interesting to note that the simple DFT calculation gives a not bad agreement with the experiment, because the height of the potential barrier turns out to be rather close to that on the CCSD(T)-corrected potential energy surface. Figure 7.15 shows the profiles of the potential energy along the instanton trajectories for the above *ab initio* methods. The variable z in this figure is the aforementioned effective coordinate along the instanton path.

It is very instructive to compare the exact expression of the pre-exponential factor B' with the approximation used by Tautermann and co-workers [42,200]. In their calculations the pre-exponential factor was taken from the one-dimensional theory by Garg [199],

$$B' = 2\omega\sqrt{\frac{m\omega a^2}{\pi}} \exp\left(\int_{-a}^{0} \left[\frac{m\omega}{p(x)} - \frac{1}{a+x}\right] dx\right), \tag{7.57}$$

where m is the effective mass, $\pm a$ are the positions of the minima of the inverted double well potential $V(x)$ along the instanton trajectory, ω is the corresponding frequency, and $p(x) = \sqrt{2mV(x)}$ is the classical momentum. Exactly the same results were derived in scope of the path-integral formalism in Reference [43]. It was further shown that in the N-dimensional case the exact expression differs from Equation (7.57) by the transverse pre-exponential factor that comes from the $N-1$ motions orthogonal to the instanton trajectory. To analyze the accuracy of Equation

TABLE 7.11
Tunneling Splitting and Instanton Parameters for the Ground State in $(DCOOH)_2$. All DFT Calculations Have Been Performed with the B3LYP Functional. The 1D Approximation Refers to the Approximate Treatment of the Pre-Exponential Factor with and without Inclusion of the Zero-Point Energy (ZPE)

method	Barrier(cm^{-1})	Δ_0(cm^{-1})	B'(cm^{-1})	S_0 (a.u.)	S_1 (a.u.)
DFT/6-31 + G(d)	2933	0.0077	3371	16.08	−3.09
with CCSD(T) correction	4515	0.00006	3371	21.0	−3.09
DFT/6-311 + +G(3df, 3pd)	2280	0.17	2697	12.7	−3.04
with CCSD(T) correction	2837	0.0038	2697	16.5	−3.04
DFT/aug-cc-pVTZ	2251	0.23	2672	12.47	−3.14
with CCSD(T) correction	2771	0.0044	2672	16.47	−3.14
DFT/6-31 + G(d) (1D approximation)		0.00003	1312	16.08	1.56
DFT/6-31 + G(d) (1D approximation + ZPE)		0.00035	1312	13.58	1.56
Experiment[a]		0.0029			

[a] Reference [201].

(7.57) we first rewrite it in the invariant form such that the calculation can be carried out using precisely the same *ab initio* data as before. One can rewrite the integral term in Equation (7.57) as

$$\int_{-a}^{0}\left[\frac{m\omega}{p(x)}-\frac{1}{a+x}\right]dx=\int_{-\infty}^{0}\left[\omega-\frac{1}{m}\frac{\partial p(x)}{\partial x}\right]d\tau+\ln\frac{p(0)}{am\omega}. \tag{7.58}$$

Inserting this into Equation (7.57) we obtain

$$B'=2\sqrt{\frac{\omega}{\pi m}}p(0)\int_{-\infty}^{0}\left[\omega-\frac{1}{m}\frac{\partial p(x)}{\partial x}\right]d\tau. \tag{7.59}$$

Finally, using the relation between the potential and momentum along the trajectory we can rewrite the above expression in the final form [see Equation (6.160)]

$$B'=2\sqrt{\frac{2\omega V(0)}{\pi}}\int_{-\infty}^{0}\left[\omega-\frac{1}{2V(\tau)}\frac{\partial V(\tau)}{\partial \tau}\right]d\tau. \tag{7.60}$$

This is equivalent to Equation (7.57), but much easier to use, since all we have to know is the potential as a function of time τ along the instanton. For the purpose of comparison we performed such "one-dimensional" calculations for the DFT/B3LYP potential with the 6-31+G(d) basis set, and the results are also shown in Table 7.11 as

$B' = 1312$ and $\Delta_0 = 0.00003$. One can see that the exponential part S_1 of the prefactor in the one-dimensional approximation is totally different, which leads to two orders of magnitude smaller tunneling splitting. The effect of the transverse modes is often taken into account in the spirit of the adiabatic path Hamiltonian by including the zero-point energy correction. This energy correction contributes to the effective potential along the tunneling path and changes S_0. However, it is still unable to reproduce the multidimensional effect completely (see $\Delta_0 = 0.000035$ in Table 7.11). Notice that the present values of the approximate instanton theory differ from those reported by Tautermann et al. [200], because those authors use an approximate tunneling path that is a combination of minimum energy and the straight-line path. Furthermore the MP2 correlation corrections are included along this tunneling path. It is important to emphasize that in terms of the parameter z the point $z = -1$ is a singular point of the equation for the matrix $\tilde{\mathbf{A}}(z)$. Mathematically, this singularity is the main reason for the substantial contribution of the transversal degrees of freedom to S_1 and this effect is totally absent in the adiabatic approximation.

In the following we calculate tunneling splittings accompanying the fundamental vibrational excitations. In particular, we focus on the comparison between the present and the adiabatic instanton approximation. For illustration purposes, we use the DFT/B3LYP[6-31+G(d)] level of quantum chemistry only. In the case of excited-state tunneling splittings the adiabatic approximation can be obtained by neglecting the left-hand side $dU_k/d\tau$ in Equation (6.191). This gives the solution $\theta_{ad}(\tau)$, which smoothly changes along the instanton trajectory, and the tunneling splitting can be approximated as

$$\frac{\Delta_{n_\gamma=1}}{\Delta_0} = \frac{\omega_\gamma}{\theta_{ad}(0)} \exp\left(2\int_{-\infty}^{0} [\omega - \theta_{ad}(\tau)]d\tau\right). \tag{7.61}$$

The behavior of $\theta_{ad}(\tau)$ at $\tau = -\infty$ is actually the most decisive factor for the ratio $\Delta_{n_\gamma=1}/\Delta_0$. This quantity has a meaning of the average energy redistributed among all the transverse degrees of freedom. In a system with strong anharmonicity $\theta(\tau)$ for high-frequency excitations rapidly decreases due to the energy exchange with low-frequency modes, leading to large positive values of ΔS_1 and exponential growth of the tunneling splitting. Such behavior has been previously found in the hydroperoxy radical HO_2 and confirmed by exact quantum calculations as reported in Section 7.2 [45]. On the other hand, in the adiabatic approximation, $\theta_{ad}(\tau)$ is close to the corresponding transversal local frequency along the instanton, which always leads to moderate values of $\Delta_{n_\gamma=1}/\Delta_0$. In this approximation, the excitation increases or deceases the tunneling splitting depending on the symmetry of the excitation mode. For the present mode the numerical analysis of Equation (6.175) shows that this simple picture completely breaks down for higher excitation energies. In Table 7.12 we compare the excited-state tunneling splitting obtained by Equations (6.175) and (6.170) with the result of the adiabatic approximation. In the present case, the longitudinal normal mode turns out to be $\gamma = 3$ with the fundamental frequency of 169 cm^{-1} (see Table 7.12).

The displacement vector of this mode is shown in Figure 7.16. The two lower frequency modes are orthogonal to the molecular plane. One can see that in the

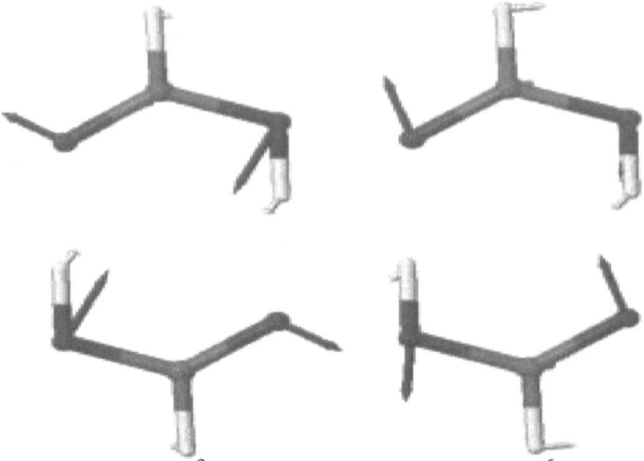

FIGURE 7.16 Displacement vectors for modes $\gamma = 4$ and $\gamma = 6$ in the case of formic acid dimer (see Table 7.12). (Taken from Reference [184] with permission.)

low-frequency region the adiabatic approximation works relatively well. In particular, we note the pronounced effect of hindering modes ($\gamma = 2, 4,$ and 6) where the excitation suppresses the tunneling effect. As an example, we show the displacement vector for mode $\gamma = 6$ in Figure 7.16. For higher frequency modes, however, the adiabatic approximation fails completely. In Figure 7.17 we compare the exact numerical solution θ_{ex} of Equation (6.191) with the adiabatic approximation θ_{ad}. As mentioned before, the starting point $z = -1$ is a singular point of Equation (6.191) that corresponds to $\tau = -\infty$ (potential minimum). In Figure 7.17(a) one can see that for the lowest excitations $\theta_{ad}(z)$ (dotted line) essentially reproduces $\theta_{ex}(z)$ (solid line). For higher frequency modes, on the other hand, the picture changes drastically [see Figure 7.17(b)]. We have found that for excitation energies higher than $\omega_\gamma = 672\ \text{cm}^{-1}$ the vibrational selectivity is lost, the extra exponential factor ΔS_1 becomes large, and the tunneling splitting grows fast with the excitation energy. It should be mentioned, however, that in this frequency range the computed splittings become much more sensitive to the accuracy of the ground-state semiclassical solution. Thus, due to finite number of *ab initio* points and inherent limited accuracy of the potential data interpolation, the matrix **A** can be computed only with two stable significant digits. For the lower excitations, the numerical accuracy of Δ_γ is similar, but for $\gamma > 7$ the inaccuracy can reach 40–100%. Therefore, the results for $\gamma = 7 - 11$ in Table 7.12 should be considered as an estimation with the correct order of magnitude. For the modes higher than $\gamma = 11$ the value of ΔS_1 becomes comparable with the classical action. This indicates the breakdown of the theory and we do not show the corresponding splitting in Table 7.12. The failure of the instanton approach for higher modes is not surprising. The main assumption of the theory is the localization of the wave function that remains harmonic in the close vicinity of the potential minimum. The fast change in $\theta(z)$ for higher modes indicates a strong anharmonicity effect so

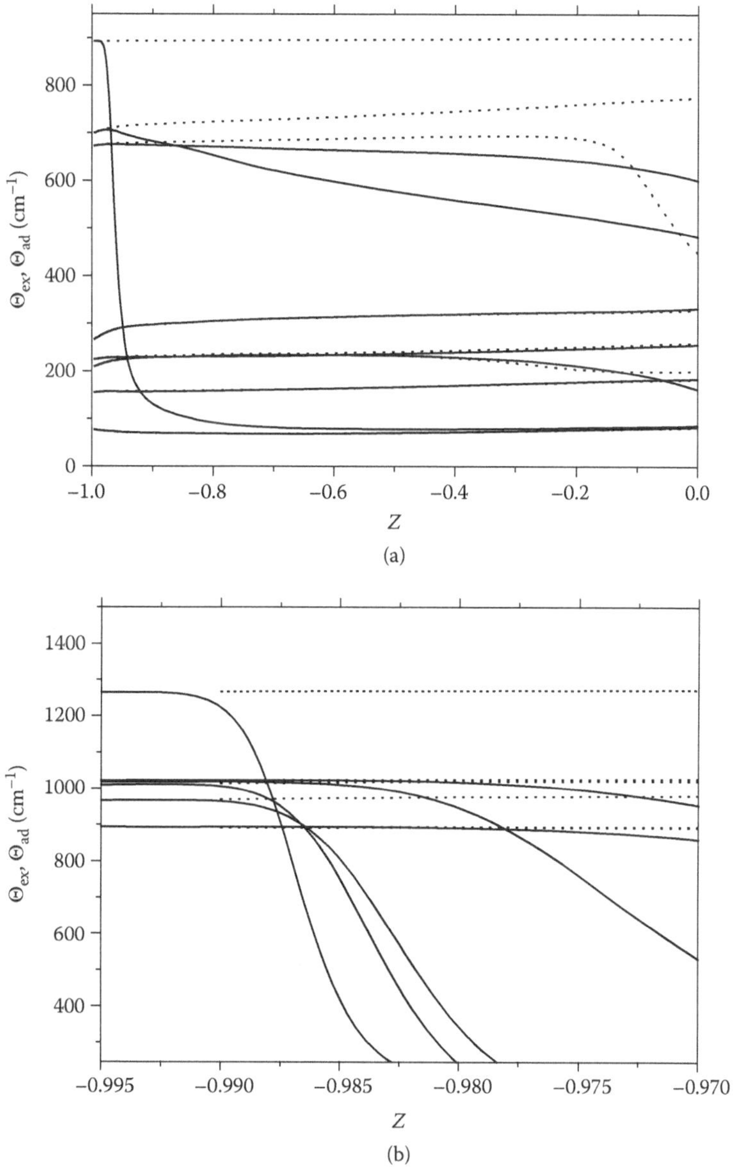

FIGURE 7.17 Effective frequency $\theta(z)$ as a function of the parameter z along the instanton path in the case of formic acid dimer. (a) Low excitation frequency region. (b) High excitation frequency region. θ_{ex} (solid line) represents the exact solution of Equation (6.191), while the adiabatic approximation θ_{ad} (dotted line) is obtained by diagonalization of the matrix **A**. (Taken from Reference [184] with permission.)

TABLE 7.12
Tunneling Splitting of the Exited States Along Modes γ in $(DCOOH)_2$ at the DFT/B3LYP-6-31+G(d) Level of Theory

γ	Freq. [ω_γ(cm^{-1})]	Δ_γ/Δ_0(exact semicl.)	Δ_γ/Δ_0(adiabatic approx.)
1	78.7	1.07	1.05
2	154	0.79	0.78
3	169	11	11
4	208	0.69	0.74
5	224	1.21	1.1
6	263	0.42	0.44
7	672	3×10^2	1.35
8	697	9×10^2	0.60
9	892	3×10^3	0.97
10	895	3.4×10^3	0.96
11	964	6.4×10^3	0.33
12	1007	...	1.15
13	1019	...	0.87
14	1022	...	0.42
15	1264	...	0.67
16	1267	...	0.83
17	1407	...	0.80
18	1443	...	0.51
19	1697	...	5.18
20	1763	...	1.77
21	2311	...	0.98
22	2314	...	0.98
23	3151	...	2.8×10^3
24	3240	...	1.5×10^3

that this assumption breaks down, which correlates with the expected chaotic behavior of the system at high energies. Unfortunately, the CO vibration, which has been used in the experiment [201] [here calculated as 1267 cm^{-1}($\gamma = 16$)] to tag the tunneling splitting, falls into this category so that no further conclusions concerning the splitting of the respective fundamental transition can be made.

8 Decay of Metastable States

8.1 GENERAL FORMULATION

In the case of decay rate problem, the potential barrier separates the potential minimum from the continuum where the system is energetically allowed to go to infinity [see Figure 1.1(b)]. The general formulation developed in Chapter 6 for tunneling splitting can be applied to the present problem with some modifications [44]. The instanton trajectory $\mathbf{q}_0(\tau)$ starts in the infinitely remote past at $\tau = -\infty$ from potential minimum and returns back to the minimum in the infinitely remote future at $\tau = \infty$. This periodic orbit always goes through a turning point on the potential contour $V(\mathbf{q}) = 0$, which can be assumed to occur at $\tau = 0$.

8.1.1 DETERMINATION OF INSTANTON TRAJECTORY

Let us recapitulate the recipe to find the instanton in the case of double well potential developed in Chapter 6. We consider the classical motion on the inverted potential with the Lagrangian [see Equation (6.107)],

$$L(\dot{\mathbf{q}}, \mathbf{q}) = \frac{1}{2}\sum_{ij} g_{ij}(\mathbf{q})\dot{q}^i\dot{q}^j + V(\mathbf{q}), \tag{8.1}$$

where $\mathbf{g}(\mathbf{q})$ is the metric tensor. We introduced the parameter $z \in [-1, 1]$ instead of the time $\tau \in (-\infty, \infty)$ and the rate of change of this parameter z for the given $\mathbf{q}_0(z)$ is given by [see Equation (6.110)]

$$\dot{z}(z) = \sqrt{\frac{2V(\mathbf{q}_0(z))}{\sum_{ij} g_{ij}(\mathbf{q}_0(z))\frac{dq_0^i(z)}{dz}\frac{dq_0^j(z)}{dz}}}. \tag{8.2}$$

By solving $dz/d\tau = \dot{z}(z)$, we can know the time-dependence of $z(\tau)$ and the trajectory $\mathbf{q}_0[z(\tau)]$ as a function of time. In order to improve the path, we use the variational principle to minimize the classical action in the space of z-parametrized paths. We look for a better instanton trajectory in the form [see Equation (6.113)]

$$q_0^i(z(\tau)) + \sum_n C^{in}\phi_n(z(\tau)) \quad \text{with} \quad \phi_n(z = \pm 1) = 0, \tag{8.3}$$

and find the coefficients $\{C^{in}\}$ by minimizing the action.

In the case of decay problem, one cannot fix both ends of the trajectory, but we can reformulate the problem in the following way. Let us try to find one-half of the instanton trajectory that runs for the semi-infinite time interval $[-\infty, 0]$ from the potential minimum to a turning point according to the classical equation of motion on the inverted potential. The corresponding boundary conditions are

$$\mathbf{q}(\tau = -\infty) = \mathbf{q}_m \text{ (potential minimum)} \tag{8.4}$$

and

$$V(\mathbf{q}(\tau = 0)) = 0. \tag{8.5}$$

Hereafter the energy is measured from the potential minimum $\mathbf{q}_m$, i.e., $V(\mathbf{q}_m) = 0$. We resort to the same strategy of iteration as before. However, since in the present case only one end of the trajectory is fixed, some modifications are required. Considering the boundary condition Equation (8.4), we choose the functions $\phi_n(z) = z^n$ for the interval $z \in [0, 1]$ with the $z \leftrightarrow \tau$ correspondence as $z(\tau = -\infty) = 0$, i.e., $\mathbf{q}_0(z = 0) = \mathbf{q}_m$. Starting with the initial guess $\mathbf{q}_0(z)$, which has only one fixed end, we can proceed in the same way as before; namely, we use the expansion Equation (8.3) and find some set of coefficients. Next, we forget about the time-dependence and redefine Equation (8.3) as $\mathbf{q}_0(z)$. This new path cannot be used as the starting point for the next iteration, however, since second boundary condition Equation (8.5) is generally violated. Thus we take one more step to renormalize the parameter z by introducing a scaling factor α so that $V[\mathbf{q}_0(\alpha z)] = 0$ is satisfied, and redefine the initial guess as $\mathbf{q}_0(\alpha z)$. Note that the shape of the path is not affected by this scaling. In other words, if the shape of the approximate instanton path is close to the correct one, it will still be so, even when $\alpha \neq 1$. Since we use the expansion

$$q_0^i(z) = q_m^i + \sum_n C^{in} z^n, \tag{8.6}$$

the scaling mentioned above simply corresponds to the transformation

$$C^{in} \to \alpha^n C^{in}. \tag{8.7}$$

Now the procedure for the calculation of instanton can be summarized as follows. Without loss of generality we set $\mathbf{q}_m = 0$.

1. At the kth step of iteration the approximate instanton *path* is supposed to be given by

$$q_0^i(z) = \sum_{n=1}^{N_b} C_{(k)}^{in} z^n \; (i = 1, 2, \ldots, N), \tag{8.8}$$

and the time-dependence of the parameter $z_{(k)}(\tau)$ is given by Equation (8.2).

2. The classical Euclidean action

$$\begin{aligned} S_E\left[\{q_0^i\left(z_{(k)}(\tau)\right) + \delta q_0^i(\tau)\}\right] &= S_E\left[\left(q_0^i\left(z_{(k)}(\tau)\right) + \sum_{n=1}^{N_b} \delta C_{(k)}^{in} z_{(k)}^n(\tau)\right)\right] \\ &\equiv S_E[\{\delta C_{(k)}^{in}\}] \end{aligned} \tag{8.9}$$

is minimized to obtain the $N \times N_b$ coefficients $\delta C_{(k)}^{in}$. The next iteration is provided by Equation (8.8) with the new coefficients,

$$C_{(k+1)}^{in} = \alpha^n \left(C_{(k)}^{in} + \delta C_{(k)}^{in} \right), \tag{8.10}$$

where α is found to satisfy the condition Equation (8.5).

3. Continue the iteration until the convergence is attained.

8.1.2 Formulation in Terms of Cartesian Coordinates

We derive the pre-exponential factor B in a convenient form in the Cartesian coordinates. The derivation is quite similar to the case of tunneling splitting. We consider the functional integral of the harmonic fluctuations around the instanton trajectory $\mathbf{q}_0(\tau)$ [see Equation (6.7)],

$$I = \int D\mathbf{q} \exp \left[- \int_{-\infty}^{\infty} \left(\frac{1}{2}\dot{\mathbf{q}}^2 + \sum_{ij} \frac{1}{2} V_{ij}''(\mathbf{q}_0(\tau) \right) q_i q_j) d\tau \right]. \tag{8.11}$$

As before in Chapter 6, V_{ij}'' is the matrix of the second derivatives of the potential $\partial^2 V/\partial q^i \partial q^j$ and the integration $\int D\mathbf{q}$ is carried out over trajectories subjected by the condition $\mathbf{q}(\pm\infty) = 0$. The functional integral diverges, since the operator $\left[-\partial^2/\partial\tau^2 + V_{ij}''(\mathbf{q}_0(\tau)) \right]$ has a zero eigenvalue. This divergence is actually one-dimensional by its nature and originated from the invariance of the classical action with respect to the arbitrary time shift, say τ_0, of the instanton $\mathbf{q}_0(\tau - \tau_0)$. It is well known that this singularity in Equation (8.11) can be transformed into the integration over τ_0 [38]. To calculate the pre-exponential factor for the multidimensional problem, we follow the same method used before and extract the one-dimensional singular part explicitly.

Let X be the coordinate along the instanton trajectory $\mathbf{q}_0[X(\tau)]$ such that $X(\pm\infty) = 0$ and let X_0 be the turning point: $X(0) = X_0$ and $\dot{X}(0) = 0$. We introduce a set of $N - 1$ transverse coordinates $\{\xi_\alpha\}$ by the relation

$$q_i = q_{0i}(X) + \sum_{\alpha} \tau_{i\alpha}(X)\xi_\alpha. \tag{8.12}$$

The orthogonal transformation matrix $\tau_{i\alpha}(X)(\alpha = 1, 2, \ldots, N-1)$ with $\tau_{iN}(X) \equiv d\mathbf{q}_{0i}/dX$ is chosen to satisfy the extra condition [see Equation (6.11)],

$$\frac{d\tau_{i\alpha}(X)}{dX} = C_\alpha(X)\tau_{iN}(X), \tag{8.13}$$

where C_α defines the curvature of the instanton trajectory. The instanton $\mathbf{q}_0(\tau)$ approaches the potential minimum along one of the normal modes, which we numerate as Nth. The local transverse coordinate ξ_α can be chosen to coincide with other $(N-1)$

normal modes at the potential minimum. Making the coordinate transformation in the functional integral, we obtain

$$I = \int Du \exp\left[-\int_{-\infty}^{\infty} L_u(u, \dot{u}, \tau) d\tau\right] \int D\{\xi\} \exp\left[-\int_{-\infty}^{\infty} L_\xi(\{\xi\}, \{\dot{\xi}\}, \tau) d\tau\right], \tag{8.14}$$

where L_u and L_ξ are quadratic Lagrangians for longitudinal and transverse fluctuations given by

$$L_u = \frac{1}{2}\dot{u}^2 + \frac{1}{2}\frac{\partial^2 V}{\partial X^2} u^2 \tag{8.15}$$

and

$$L_\xi = \frac{1}{2}\sum_\alpha \dot{\xi}_\alpha^2 + \frac{1}{2}\sum_{\alpha\beta}\left[\frac{\partial^2 V}{\partial \xi_\alpha \partial \xi_\beta} - 3\dot{X}^2 C_\alpha C_\beta\right]\xi_\alpha \xi_\beta, \tag{8.16}$$

where the second term in the bracket comes from the curvature of $\mathbf{q}_0(\tau)$. All the quantities in these equations are taken on the instanton trajectory; thus, $\dot{X} \equiv \dot{X}(\tau)$, $C_\alpha \equiv C_\alpha[X(\tau)]$, and $\partial^2 V/\partial S^2 \equiv \partial^2 V/\partial S^2[\mathbf{q} = \mathbf{q}_0(\tau)]$, where $S = \xi_\alpha$ or X. The representation Equation (8.14) allows us to extract the zero-mode singularity exactly in the way as in the one-dimensional case. The same derivation used before gives the pre-exponential factor in the form [see Equation (6.26) and the discussions there]

$$B = \sqrt{\frac{2S_0}{\pi}}\left[\frac{\det'\left(-\frac{\partial^2}{\partial \tau^2} + \frac{\partial^2 V}{\partial X^2}(\tau)\right)}{\det\left(-\frac{\partial^2}{\partial \tau^2} + \omega_N^2\right)}\right]^{-1/2}\left[\frac{\det\left(-\frac{\partial^2}{\partial \tau^2}\mathbf{1} + \mathbf{\Omega}^2(\tau)\right)}{\det\left(-\frac{\partial^2}{\partial \tau^2}\mathbf{1} + \mathbf{\Omega}_0^2\right)}\right]^{-1/2}. \tag{8.17}$$

Here det[$\cdots$] is understood as the infinite product of the eigenvalues of the corresponding differential operator with the Dirichlet boundary condition at $\tau \to \pm\infty$. The prime in the first numerator indicates that the zero eigenvalue of $[-\partial^2/\partial\tau^2 + V''_{XX}]$ is omitted. We also introduce the $(N-1)$-dimensional symmetric matrices $\mathbf{\Omega}(\tau)$ and $\mathbf{\Omega}_0$ defined by

$$\Omega^2_{\alpha\beta}(\tau) = \frac{\partial^2 V}{\partial\alpha\partial\beta} - 3\dot{X}^2 C_\alpha C_\beta \tag{8.18}$$

and

$$(\Omega_0)_{\alpha\beta} = \delta_{\alpha\beta}\omega_\alpha, \tag{8.19}$$

where ω_α are the $N-1$ transverse normal frequencies at the potential minimum. The transverse part of the pre-exponential factor [second ratio of determinants or the last factor in Equation (8.17)] has been already evaluated before. The derivation was based only on the symmetry of $\Omega(\tau)$ with respect to the time inversion, and in the present case this symmetry is guaranteed by the condition $\mathbf{q}_0(\tau) = \mathbf{q}_0(-\tau)$. The results read [see Equation (6.64)]

$$\left[\frac{\det\left(-\frac{\partial^2}{\partial \tau^2}\mathbf{1} + \mathbf{\Omega}^2(\tau)\right)}{\det\left(-\frac{\partial^2}{\partial \tau^2}\mathbf{1} + \mathbf{\Omega}_0^2\right)}\right]^{-1/2} = \sqrt{\frac{\det \mathbf{\Omega}_0}{\det \mathbf{\Xi}(0)}} \exp\left[\int_{-\infty}^{0} \mathrm{Tr}\,(\mathbf{\Omega}_0 - \mathbf{\Xi}(\tau))\, d\tau\right], \tag{8.20}$$

where $(N-1) \times (N-1)$ symmetric matrix $\boldsymbol{\Xi}(\tau)$ is the solution of the equation [see Equation (6.60)]

$$\dot{\boldsymbol{\Xi}} + \boldsymbol{\Xi}^2 = \boldsymbol{\Omega}^2(\tau) \tag{8.21}$$

under the initial condition

$$\boldsymbol{\Xi}(-\infty) = \boldsymbol{\Omega}_0 = \boldsymbol{\Omega}(-\infty). \tag{8.22}$$

The longitudinal one-dimensional factor [first ratio of determinants in Equation (8.17)] has been evaluated by Shulman [205] for the quartic potential. The analogous method can be used in the general case to lead to [see also Equations (2.108), (2.109), and (2.129)]

$$\left[\frac{\det'\left(-\frac{\partial^2}{\partial\tau^2} + \frac{\partial^2 V}{\partial X^2}(\tau)\right)}{\det\left(-\frac{\partial^2}{\partial\tau^2} + \omega_N^2\right)}\right]^{-1/2} = P\omega_N\sqrt{\frac{2\omega_N}{S_0}}, \tag{8.23}$$

where the constant P characterizes the asymptotic behavior of the symmetric instanton trajectory,

$$X(\tau) \overset{\tau\to\pm\infty}{\longrightarrow} P\exp(-\omega_N|\tau|), \tag{8.24}$$

and can be easily found from the classical mechanics. The time-dependence $X(\tau)$ of the instanton trajectory is given by

$$\tau = \int^X \frac{dx}{p_0(x)} = \int_0^X \left(\frac{1}{p_0(x)} - \frac{1}{\omega_N x}\right) dx + \frac{1}{\omega_N}\ln X + C, \tag{8.25}$$

where $p_0(X) = \sqrt{2V(\mathbf{q}_0(X))}$ is the absolute value of the classical momentum. The constant of integration C is determined from the condition $X(0) = X_0$ and we have

$$P = \exp(-\omega_N C) = X_0 \exp\left(\int_0^{X_0} dx \left[\frac{\omega_N}{p_0(x)} - \frac{1}{x}\right]\right). \tag{8.26}$$

Putting all the terms together, we finally obtain the decay rate as

$$k = \omega_N X_0 \sqrt{\frac{\omega_N}{\pi}} \exp\left(\int_0^{X_0}\left[\frac{\omega_N}{p_0} - \frac{1}{x}\right] dx\right)$$
$$\times \sqrt{\frac{\det \boldsymbol{\Omega}_0}{\det \boldsymbol{\Xi}(0)}} \exp\left(\int_{-\infty}^{0} \mathrm{Tr}(\boldsymbol{\Omega}_0 - \boldsymbol{\Xi}(\tau))d\tau\right) \exp(-S_0). \tag{8.27}$$

It should be noted that S_0 is the action along the instanton from $\tau = -\infty$ to $\tau = \infty$ and thus is two times of the action from the potential minimum to the turning point.

8.1.3 General Canonically Invariant Formulation

The result of the preceding subsection is not very convenient for practical use, because it requires the solution of the differential equation Equation (8.21) that contains the Hessian of the potential in terms of the transverse local coordinates together with the corresponding curvatures. One can obtain a more useful canonically invariant form that does not rely on any local coordinates. To do this, we first rewrite the first integral in Equation (8.27) in terms of the time τ along the instanton trajectory. Using the transformation

$$\int_0^X \left(\frac{\omega_N}{p_0} - \frac{1}{x}\right) dx = \ln \frac{p_0(X)}{\omega_N X} + \int_{-\infty}^{\tau} \left(\omega_N - \frac{dp}{dx}\right) d\tau \tag{8.28}$$

and taking the limit $\tau \to 0 (X \to X_0)$, we have

$$k = \sqrt{\frac{\omega_N}{\pi}} p(0) \exp\left(\int_{-\infty}^{0} \left[\omega_N - \frac{dp}{dx}\right] d\tau\right)$$
$$\times \sqrt{\frac{\det \boldsymbol{\Omega}_0}{\det \Xi(0)}} \exp\left(\int_{-\infty}^{0} \mathrm{Tr}(\boldsymbol{\Omega}_0 - \Xi(\tau)) d\tau\right) \exp(-S_0), \tag{8.29}$$

where $\tau = 0$ should be understood as the limit $\tau \to 0^-$. The existence of this limit is evident from the previous representation for k. For $\tau < 0$ there is one-to-one correspondence $X \leftrightarrow \tau$ and we can introduce the function

$$W_0(\mathbf{q}) = \int_0^X p_0(X) dX + \frac{1}{2} \sum_{\alpha\beta} \Xi_{\alpha\beta}(X) \xi_\alpha \xi_\beta, \tag{8.30}$$

which, up to the terms $O(\xi^3)$, satisfies the Hamilton-Jacobi equation

$$\frac{1}{2}\left(\frac{\partial W_0}{\partial \mathbf{q}}\right)^2 = V(\mathbf{q}). \tag{8.31}$$

This is readily checked by making the coordinate transformation Equation (8.12) and using the properties of $\{\tau_{i\alpha}\}$. The quantity W_0 has the simple physical meaning as the main exponent factor of the semiclassical wave function $\Psi \simeq \exp[-W_0(\mathbf{q})]$, which coincides with the ground state of the harmonic oscillator in the potential well. In the same way as before, we introduce the second derivative matrix

$$A_{ij}(\tau) \equiv \frac{\partial^2 W_0}{\partial q_i \partial q_j} (\mathbf{q}_0(\tau)), \tag{8.32}$$

which satisfies the more handy equation obtained by differentiating Equation (8.31) twice,

$$\frac{d}{d\tau} A_{ij} + (\mathbf{A}^2)_{ij} = \frac{\partial^2 V(\mathbf{q})}{\partial q_i \partial q_j} |_{\mathbf{q}=\mathbf{q}_0(\tau)}. \tag{8.33}$$

The initial condition follows from the existence of the limit $\mathbf{A}_m = \mathbf{A}(\tau \to -\infty)$ and reads

$$\mathbf{A}_m^2 = \frac{\partial^2 V(\mathbf{q})}{\partial \mathbf{q} \partial \mathbf{q}}|_{\mathbf{q}=\mathbf{q}_m}. \tag{8.34}$$

Now we can express $dp/dX = \partial^2 W_0/\partial X^2$ and $\Xi_{\alpha\beta} = \partial^2 W_0/\partial \xi_\alpha \partial \xi_\beta$ in Equation (8.29) in terms of $\mathbf{A}$ and get rid of the local coordinates. The two terms in the integrals in Equation (8.29) are combined together as

$$\frac{dp(X)}{dX} + \sum_\alpha \Xi_{\alpha\alpha} = \frac{\partial^2 W_0}{\partial X^2} + \sum_\alpha \frac{\partial^2}{\partial \xi_\alpha^2} = \sum_i \frac{\partial^2 W_0}{\partial q_i^2} = \mathrm{Tr}\mathbf{A}. \tag{8.35}$$

At the potential minimum we have $\omega_N \Pi \omega_\alpha = \det \mathbf{A}_m$ and $\omega_N + \sum \omega_\alpha = \mathrm{Tr}\mathbf{A}_m$. Equations (8.12) and (8.30) show that Ξ is directly related to $\mathbf{A}$ as

$$\Xi_{\alpha\beta} = \sum_{ij=1}^{N} A_{ij} \tau_{i\alpha} \tau_{j\beta} \; (\alpha, \beta = 1, 2, \ldots, N-1). \tag{8.36}$$

To proceed further it is convenient to rotate the frame of reference in such a way that $\tau_{iN}(X_0) = \delta_{iN}$. This simply means that the coordinate q_N is chosen along the tangent to the instanton trajectory at the turning point. Then, $\det \Xi$ is equal to the minor of the (N, N) element of $\mathbf{A}$ and we obtain [cf. Equation (6.80)]

$$\det \Xi = \det \mathbf{A} \sum_{ij} (\tau_{iN} A_{ij}^{-1} \tau_{jN}). \tag{8.37}$$

Using Equations (8.35) and (8.37) and the fact that τ_{iN} are the components of the unit tangent vector along the instanton trajectory, we finally obtain the decay rate as

$$k = \sqrt{\frac{\det \mathbf{A}_m}{\pi \det \mathbf{A}(0)}} \frac{\mathbf{p}^2(0)}{\sqrt{\sum_{ij} p_i(0) A_{ij}^{-1}(0) p_j(0)}}$$

$$\times \exp\left(\int_{-\infty}^{0} d\tau [\mathrm{Tr}(\mathbf{A}_m - \mathbf{A}(\tau))] \right) \exp(-S_0). \tag{8.38}$$

This result is valid for arbitrary choice of coordinates as it is invariant under rotation of the frame of reference.

It should be recalled that the zero value of the time argument in $\mathbf{p}(0)$ and $\int_{-\infty}^{0} \cdots d\tau$ should be understood as the limit $\tau \to 0^-$. Except for this peculiarity and the factor 2, Equation (8.38) looks identical to the canonically invariant expression for the tunneling splitting derived before [see Equation (6.81)]. Indeed, there is close analogy between the two problems. We have shown before that the instanton theory is equivalent to the ordinary WKB approximation to the Schrödinger equation. In the case of double well potential, the tunneling splitting can be evaluated by the Herring formula with use of the WKB wave function. The similar type of WKB approach is possible in the present problem and the decay rate can be evaluated from the flux at the outlet to the continuum where the wave function is obtained by a sort of analytical continuation. This approach does not lead to any extra difficulties and Equation (8.38) for the decay rate can be rederived in the scope of the WKB approximation in the same way as before. Schmid [33] used this method to solve the two-dimensional problem. For this particular case his result coincides with Equation (8.27) above. The equation for the transverse fluctuations differs from the present one by the coefficient of the curvature term, which is not correct in Reference [33]. The equivalence of the instanton and WKB approaches is not just a matter of methodological interest, but also allows one to generalize the theory to non-Euclidean metric tensor exactly in the same way as in the problem of tunneling splitting. Due to the complete mathematical equivalence of the two problems, we present the general formula without derivation.

Let $\mathbf{q} = (q^1, q^2, \ldots, q^N)$ be coordinates on a manifold M with Riemannian metric $g_{ij}(\mathbf{q})$. We consider the Hamiltonian

$$\hat{H} = \hat{K} + \hat{V} = -\frac{1}{2\sqrt{g}} \sum_{ij} \frac{\partial}{\partial q^i} \left(\sqrt{g} g^{ij} \frac{\partial}{\partial q^j} \right) + V(\mathbf{q}), \tag{8.39}$$

where, as usual, $g = \det g_{ij}$ and the contravariant metric tensor g^{ij} is the inverse of g_{ij}. The corresponding classical Hamiltonian with the upside-down potential reads

$$H(\mathbf{p}, \mathbf{q}) = \frac{1}{2} \sum_{ij} g^{ij} p_i p_j - V(\mathbf{q}) \tag{8.40}$$

and the decay rate is finally given by the general formula

$$k = \sqrt{\frac{g_0 \det \mathbf{A}_m}{\pi g_m \det \mathbf{A}_0}} \frac{(\mathbf{p}^T \mathbf{g} \mathbf{p})_0}{\sqrt{(\mathbf{p}^T \mathbf{A}^{-1} \mathbf{p})_0}} \exp(-S_0 - S_1), \tag{8.41}$$

where

$$S_1 = \int_{-\infty}^{0} d\tau \left[\mathrm{Tr}(\mathbf{A}(\tau) - \mathbf{A}_m) + \mathbf{p}^T \Lambda \right], \tag{8.42}$$

$$\Lambda^i(\mathbf{q}) = \frac{1}{\sqrt{g}} \frac{\partial}{\partial q^j} (\sqrt{g} g^{ij}), \tag{8.43}$$

$$\mathrm{Tr}(\mathbf{A}) \equiv \sum_i A_i^i = \sum_{ij} g^{ij} A_{ji}, \tag{8.44}$$

and S_0 is the action integral along the instanton. The symmetric matrix $\mathbf{A}(\tau)$ is the solution of the equation [see also Equations (6.97) and (6.181)]

$$\dot{\mathbf{A}} = -H_{\mathbf{qq}} - H_{\mathbf{qp}} \mathbf{A} - \mathbf{A} H_{\mathbf{pq}} - \mathbf{A} H_{\mathbf{pp}} \mathbf{A}, \tag{8.45}$$

where $H_{\mathbf{qq}}, H_{\mathbf{qp}}, \ldots$ are the matrices of the second derivatives $\partial^2 H / \partial q^i \partial q^j$, $\partial^2 H / \partial q^i \partial p_j, \ldots$ on the instanton trajectory $\mathbf{q}_0(\tau)$. The indices m and 0 in Equation (8.41) indicate that the corresponding quantities are taken at the potential minimum $\mathbf{q}_m = \mathbf{q}_0(\tau = -\infty)$ and the turning point $\mathbf{q}_0 = \mathbf{q}_0(\tau = 0)$, respectively.

Finally, we should pay attention to the removable singularities in Equation (8.41) arising from both ends ($z = 0$ and 1) of the instanton path, and derive the expression suitable for actual numerical evaluation. By making use of the analytical form of the instanton path, we rewrite Equation (8.45) in terms of the parameter z,

$$\frac{d\mathbf{A}}{dz} = \frac{1}{\dot{z}(z)}[\text{RHS of Equation (8.45)}]. \tag{8.46}$$

The solution must be propagated from $z = 0(\tau = -\infty)$ to $z = 1(\tau = 0)$. Since the time derivative of the parameter z at the potential minimum is zero $\dot{z}(z = 0) = 0$, Equation (8.46) has to be integrated from a certain small initial value $z = \epsilon$. The initial condition $\mathbf{A}(\epsilon)$ is determined by the linearization of Equation (8.46). Correspondingly, one can rewrite S_1 in terms of z as

$$S_1 = \int_0^1 I(z)dz \text{ with } I(z) \equiv \frac{1}{\dot{z}(z)}\left[\text{Tr}\left(\mathbf{A}(z) - \mathbf{A}_m\right) + \mathbf{p}^T \Lambda\right] \tag{8.47}$$

and accurately determine the integrand $I(z)$ at $z = 0$. In this part the whole procedure is exactly the same as in the case of tunneling splitting. Another problem is related with taking the limit $z \to 1$, since A_{ij} diverges as $(1 - z)^{-1/2}$ and $\dot{z}(z)$ goes to zero as $(1 - z)^{1/2}$ there. The divergence of $\mathbf{A}_0 = \mathbf{A}(z = 1)$ in the pre-exponential factor of Equation (8.41) is easily understood to be removable from the expression. Actually, one can simply stop to integrate Equation (8.46) shortly before $z = 1$ and use the obtained value as $\sim \mathbf{A}(1)$. The more complicated problem is related to the divergence of S_1 itself, since $[1/\dot{z}(z)]\text{Tr}\mathbf{A} \sim [1/(1 - z)]$ and the integral in Equation (8.47) diverges logarithmically. Formally, this is exactly canceled by $\mathbf{p}_0 = \mathbf{p}(z = 1) = 0$. The origin of this problem is the transformation of Equation (8.28), which is needed to combine the transverse and longitudinal prefactors and to obtain the invariant expression of k. Noting that in scope of our computational method the parameter z itself can be treated as an invariant dimensionless coordinate along the instanton, we can remedy this problem by simply making the transformation reverse to Equation (8.28) in terms of the coordinate z. Along the instanton trajectory the momentum is given by

$$p_i(\tau) = \sum_j g_{ij}(\mathbf{q}_0(\tau))\dot{q}^j(z). \tag{8.48}$$

Using the explicit z dependence, we can rewrite $\mathbf{p}$ as

$$p_i(z) = \sum_j g_{ij}\frac{dq^j}{dz}\dot{z}(z) \equiv \tilde{p}_i(z)\dot{z}(z), \tag{8.49}$$

where the newly introduced vector $\tilde{\mathbf{p}}$ does not become zero anywhere. Correspondingly, we can rewrite the decay rate in the form

$$k = \lim_{\epsilon \to 0^+} \cdots \dot{z}(1 - \epsilon)\exp\left(-S_0 - \int_0^{1-\epsilon} I(z)dz\right), \tag{8.50}$$

where "$\cdots$" stands for the rest of the prefactor with $\mathbf{p}$ substituted by $\tilde{\mathbf{p}}$. Now, we make use of the transformation

$$\dot{z}(z) = \omega_z z \exp\left(\int_0^z dz\left[\frac{1}{\dot{z}}\frac{d\dot{z}}{dz} - \frac{1}{z}\right]\right) \tag{8.51}$$

and finally obtain

$$k = \sqrt{\frac{g_0 \det \mathbf{A}_m}{\pi g_m \det \mathbf{A}_0}} \frac{(\tilde{\mathbf{p}}^T \mathbf{g} \tilde{\mathbf{p}})_0}{\sqrt{(\tilde{\mathbf{p}}^T \mathbf{A}^{-1} \tilde{\mathbf{p}})_0}} \omega_z \exp\left(-S_0 - \int_0^1 dz \left[I(z) - \frac{1}{\dot{z}} \frac{d\dot{z}}{dz} + \frac{1}{z}\right]\right), \tag{8.52}$$

where, by definition, $\omega_z = \dot{z}/z|_{z \to 0}$. It should be noted that the suffix 0 corresponds to $\tau = 0$ or $z = 1$ [see Equation (8.41) and the explanation below Equation (8.45)]. Once the trajectory is found, i.e., the coefficients $\{C^{in}\}$ are determined, all the quantities in Equation (8.52) are calculated easily, including ω_z. The integral term is convergent now and can be evaluated with an appropriate Gaussian quadrature. Equation (8.52) is the final practical formula, which, together with the recipe for finding the instanton trajectory, gives a numerically very stable procedure to calculate the decay rate.

8.2 NUMERICAL APPLICATION

To demonstrate the applicability of the theory, we consider the particle of mass m in the N-dimensional cubic potential

$$V(\mathbf{x}) = \frac{m\omega_0^2 x_1^2}{2}\left(1 - \frac{x_1}{x_0}\right) + \sum_{i=2}^{N}\left(\frac{m\omega_i^2 x_i^2}{2} + d_i x_i x_1\right). \tag{8.53}$$

First we check the iterative procedure for the instanton trajectory. We set $x_0 = m = \omega_i = 1$ and calculate the instanton for different coupling constants d_i. The test has confirmed the excellent convergence for the dimension N ranging from 2 to 20 and the coupling constant $d_i \in [0, 1]$. In all these cases, 10 ($= N_b$) basis functions were enough to get five significant digits in the classical action S_0. For $N_b = 30$ the accuracy increased up to six significant digits. The initial guess for the instanton path was taken as the straight line connecting the point $\mathbf{x} = 0$ and a certain point on the equipotential $V(\mathbf{x}) = 0$. The number of iterations N_{iter} strongly depends on this guess, as shown in Table 8.1. The stronger the deviation of the initial guess from the true

TABLE 8.1
Dependence of the Number of Iterations N_{iter} on the Initial Guess of the Instanton Trajectory in the Case of the Potential of Figure 8.1. θ Is the Angle between the Initial Path and X_1 Axis. S_{init} Is the Classical Action for the Initial Guess. The Convergent Value Is $S_0 = 0.22205$

θ	$(S_0)_{\text{init}}$	N_{iter}
0.1	0.27	4
0.2	0.38	5
0.3	0.7	9
0.4	1.7	17
0.5	5.6	40

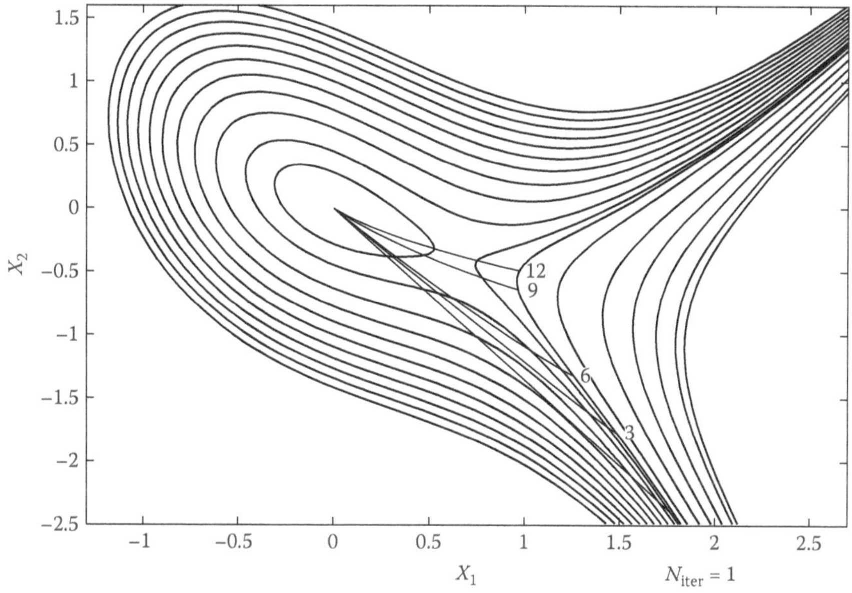

FIGURE 8.1 Iteration process of the calculation of the instanton trajectory in the cubic potential for $N = 2$ and $d_2 = 0.6$. N_{iter} is the number of iterations. (Taken from Reference [44] with permission.)

instanton is, the larger the value of the initial action $(S_0)_{\text{init}}$ is and the more iterations are required. Figure 8.1 shows the iterations for $N = 2$ and $d_2 = 0.6$. The "free end" of the trajectory is seen to slide along the equipotential $V = 0$ in the course of iteration. The initial guess on Figure 8.1 corresponds to $\theta = 0.4$ and $(S_0)_{\text{init}} \simeq 1.7$, where θ is the angle between the initial path and the X_1 axis. The converged value of the classical action is $S_0 = 0.22205$.

To test the whole numerical algorithm and accuracy of the decay rate calculation, we employ the nonlinearly transformed model of the separable case with $d_i = 0$. Thus, the correct instanton trajectory is just a straight line $x_i = 0 (i = 2, \ldots, N)$ and the rate k is simply given by Equation (8.27) without the transverse part of the prefactor $\left(= \sqrt{\frac{\det \mathbf{\Omega}_0}{\det \mathbf{\Xi}(0)}} \exp\left(\int_{-\infty}^{0} \text{Tr}(\mathbf{\Omega}_0 - \mathbf{\Xi}(\tau)) d\tau\right)\right)$. The integrals can be performed analytically to give the well-known result given by Equations (2.146) and (2.147). The nonlinear transformation $\mathbf{x} \rightarrow \mathbf{q}$ used is the same as Equation (7.2),

$$q^k = T_{k1} x_1 + \sum_{i=2}^{N} T_{ki} (x_i + 1 + x_1^2)^2 \; (k = 1, 2, \ldots, N), \tag{8.54}$$

where T_{ki} is an orthogonal matrix in the form

$$T = T^{(1n)} T^{[1(n-1)]} \cdots T^{(12)} \tag{8.55}$$

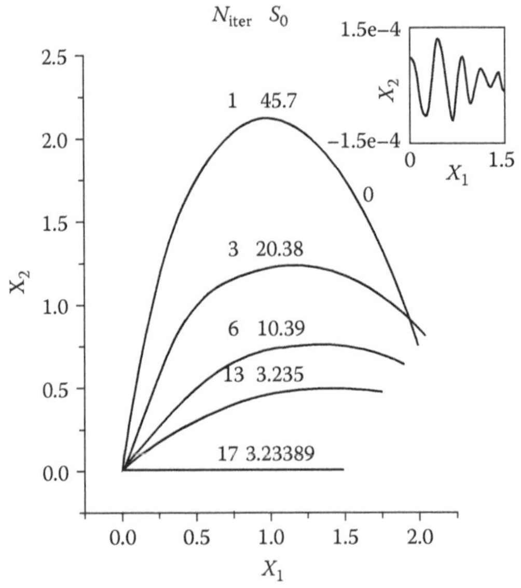

FIGURE 8.2 Iteration process for the instanton trajectory in the separable 20-dimensional cubic potential model. The calculation is carried out within the framework of the nonlinearly transformed coordinates [see Equations (8.54) and (8.55)]. N_{iter} is the number of iterations and S_0 is the classical action. Its exact value is $S_0 = 3.233859$. The inset shows the enlarged scale of the converged trajectory after 17 iterations. (Taken from Reference [44] with permission.)

and $T^{(ij)}$ is the matrix of rotation by angle γ in (x_i, x_j) plane. In the new coordinates $\mathbf{q}$ the Hamiltonian operator has the form of Equation (8.39) and it is readily checked that on the instanton trajectory the Jacobian of the transformation is not zero. The final formula is invariant under arbitrary coordinate transformation. At the same time the computation in terms of the new coordinates $\mathbf{q}$ allows one to check the numerical procedure by comparison with the analytical result as in Section 7.1. The parameters used are $\gamma = 0.1$, $\omega_i = 1000$ cm^{-1}, $\omega_0 = 650$ cm^{-1}, $x_0 = 1.5$ a.u., and m is the proton mass. The calculations were carried out for $N = 2, 4, 10, 20$. We found the accuracy to be the same in all cases and present the result only for $N = 20$ and $N_b = 10$. Figure 8.2 shows the iterative calculations of the instanton trajectory. The initial guess was taken as the straight line in q space, i.e., $q^i = 0 (i \neq 1)$, connecting the potential minimum and the equipotential $V = 0$. For better illustration, the back transformation $\{\mathbf{q}\} \rightarrow \{\mathbf{x}\}$ is made and the shape of the trajectory in the (x_1, x_2) plane is shown. After 17 iterations with $N_b = 10$ the trajectory converges to the straight line in x space, i.e., $x_i = 0 (i \neq 1)$, and the accuracy of the action is five significant digits.

In Figure 8.3 numerically evaluated $I(z)$ is compared with the exact analytical value in the present model. Some disagreement in a small area can be seen, which is related to the linear approximation at $z = 0$ mentioned above. The stable three digits of accuracy are attained in the decay rate and the final value obtained is $k = 3.611 \times 10^{-5}$ in comparison with the analytical answer $k_{exact} = 3.608 \times 10^{-5}$.

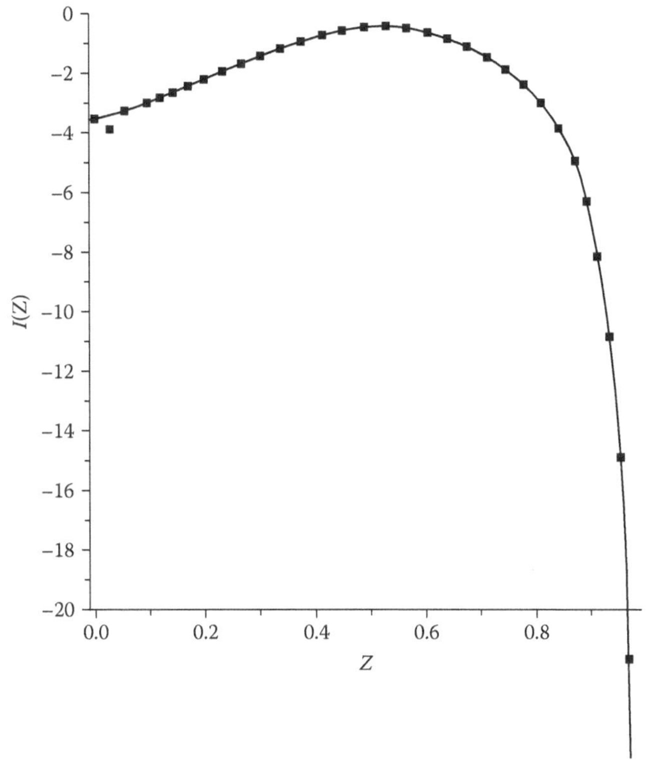

FIGURE 8.3 Behavior of the integrand $I(z)$ in the case of cubic potential. Solid squares represent the numerical results obtained by solving the matrix equation, Equation (8.46). Solid line is the exact $I(z)$. (Taken from Reference [44] with permission.)

9 Tunneling in Chemical Reactions

9.1 DETERMINATION OF CAUSTICS AND PROPAGATION IN TUNNELING REGION

In the case of tunneling in chemical reactions [see figure1.1(c)], the total energy of reagents cannot be generally considered to be small; treating it as the first-order term of $\hbar$ as was successfully done in Chapters 6 and 7 does not lead to any sensible semiclassical theory. The semiclassical mechanics requires (1) construction of Lagrange manifolds, (2) determination of a multidimensional caustic surface, and (3) connection of wave functions between classically allowed and forbidden regions [59]. Unfortunately, these problems have not yet been solved mathematically. As was discussed in Section 3.1, in the framework of the ordinary WKB approach we can construct Lagrange manifolds and caustics in the case of two dimensions and the connection problem can be treated relatively well by assuming the local separability. Extension of this method to higher dimensions, however, cannot be practical. The complex-valued (complex coordinate and/or momentum) WKB theory has been devised, as briefly explained there, but cannot be simply extended to multidimensions (see, for instance, [15]). This formidable task can hardly be accomplished mainly because of the complexity of the Lagrange manifold and the fact that analytical properties of any realistic potential energy surfaces extended into the complex-valued coordinate space cannot be guaranteed. The use of a complex trajectory was nicely materialized by the double-ended complex trajectory in the classical S-matrix theory [206], in which the quantized boundary conditions are satisfied at both (initial and final) ends. Unfortunately, this method turned out not to be practically useful, because finding such trajectories is also a formidable task in a multidimensional complex coordinate (and/or momentum) space. We can, however, make a break through the difficulty of detecting caustics in multidimensional space. An efficient way of doing that has been devised along classical trajectory [47]. This is not a trivial task at all, since it is not possible to find caustics in high-dimensional space by just shooting classical trajectories. Since the caustics can be found along each classical trajectory, chemical dynamics with tunneling effects incorporated can now be treated by the on-the-fly method. In this subsection, this method is explained.

It is known that in a $2N$-dimensional phase space, the N-dimensional Lagrange manifold can be generated by a bunch of classical trajectories $[\mathbf{p}(\alpha), \mathbf{q}(\alpha)]$ parametrized by an N-dimensional parameter $\alpha = t, \alpha_1, \alpha_2, \ldots, \alpha_{N-1}$, where t is time and the rest of $(N-1)$ parameters determine the so-called initial Lagrange manifold, which must be chosen according to the specified integrals of motion (the classical analog of quantum numbers). The projection of this manifold onto the configuration space $[\mathbf{q}(\alpha)]$ is generally not diffeomorphic; namely, the singularity occurs at the caustics where the determinant $|\frac{D\mathbf{q}}{D\alpha}|$ vanishes (=0). Assuming that there is no caustics initially, one can rewrite the above condition in the form $|\frac{D\mathbf{q}(t)}{D\mathbf{q}(0)}| = 0$. Being considered as a function of t, the stability matrix $|\frac{D\mathbf{q}(t)}{D\mathbf{q}(0)}|$ satisfies the second-order matrix differential equation for linear deviations of a classical trajectory [207,208]. For long propagation times, the solution of such an equation may become unstable due to exponentially growing and decaying terms, which can be several orders of magnitude apart. An equivalent form of the singularity condition is $\mathbf{A} \equiv |\frac{\mathbf{DP}}{\mathbf{Dq}}| = \infty$. This matrix $\mathbf{A}$ satisfies a more handy Riccati-type differential equation [43]:

$$\dot{A} = -H_{\mathbf{qq}} - H_{\mathbf{qp}}A - AH_{\mathbf{pq}} - AH_{\mathbf{pp}}A, \tag{9.1}$$

where $H_{\mathbf{qq}}, H_{\mathbf{qp}} \cdots$ are the matrices of second derivative of the classical Hamiltonian taken along the trajectory, i.e., $\partial^2 H/\partial q\partial q, \partial^2 H/\partial q\partial p, \cdots$. This is the same equation we have already met in Chapter 6 [see Equation (6.97)]. The higher degree of numerical stability common to this type of nonlinear differential equation makes this approach particularly appealing, but the divergence of the solution at the caustics persists and hinders solution of this equation beyond the singularity. This divergence is an intrinsic property of the solution and cannot be avoided simply by improving the sophistication of the numerical recipe. By this reckoning, we can use a sequence of canonical transformations capable of avoiding the divergence of the solution of the differential equation while accurately locating the caustics. During propagation of the classical trajectory, the solution of Equation (9.1) diverges at the caustics where

$$\det |A(t_{caustics})| = \infty. \tag{9.2}$$

As stated above, further solution of the differential equation is not possible beyond this point and we have to reformulate the problem. To achieve this we use a certain transformation. To clarify the idea, let us first consider a one-dimensional problem in which the caustics just becomes the turning point q_0. As the trajectory reaches q_0, the momentum $p(q) \sim \sqrt{q - q_0}$ becomes zero and $A = dp/dq \sim 1/\sqrt{q - q_0}$ diverges, which eventually breaks down any computational scheme near q_0. At the same time the inverse A^{-1} remains finite and it can be safely propagated beyond the turning point by a similar Riccati-type differential equation that can be obtained from Equation (9.1). We further notice that the coordinate and momentum can be exchanged by the classical canonical transformation $(p, q) \rightarrow (\tilde{p}, \tilde{q}) \equiv (q, -p)$ so that A^{-1} coincides, up to the sign, with $\tilde{A} = d\tilde{p}/d\tilde{q}$ in terms of the new canonical variables. The sign change ensures that the transformation is canonical and that Hamilton's equations of motion remain invariant in the new representation. It is relevant to reiterate that the second-derivative coefficients are also changed in the new representation and it can be shown that Equation (9.1) does not change under the transformation. In the N-dimensional case ($N > 1$), the naive approach would be to invert the entire A matrix as in the one-dimensional case; but this turns out to be ineffective. The reason is that if the matrix A has an element that is a derivative of an integral-of-motion, such an element vanishes and the inversion of the matrix diverges for reasons not directly related to caustics. A more workable approach is to selectively invert only the diverging element(s) of the old matrix. If the diverging element is assumed to be A_{NN}, the inversion of this element corresponds to a transformation of the type

$(p_N, q_N) \rightarrow (\tilde{p}_N, \tilde{q}_N) = (q_N, -p_N)$, which mixes the momenta and coordinates in the new representation. The new matrix $\tilde{A}$ becomes free of any divergent element and its propagation proceeds smoothly through caustics.

For numerical implementation, the sequential steps to carry out these procedures are summarized as follows:

1. For an $N \times N$ matrix, irrespective of the position of the diverging matrix element, a simple rotation ensures the repositioning of the diverging element as the (N, N) element of the matrix. This rotation is a canonical transformation and we note that such a rotation is best achieved using the orthogonal matrix which diagonalizes A. Writing this diagonalizing matrix T, the rotationally transformed version of A is obtained by

$$p' = Tp \text{ and } q' = Tq \tag{9.3}$$

so that

$$A' = TAT^t. \tag{9.4}$$

2. By invoking the transformation $(p'_N, q'_N) \rightarrow (-\tilde{q}_N, \tilde{p}_N)$, we obtain $A' \rightarrow \tilde{A}$, i.e.,

$$\frac{D(p')}{D(q')} \rightarrow \frac{D(\tilde{p})}{D(\tilde{q})}, \tag{9.5}$$

where

$$\tilde{p}_1, \ldots, \tilde{p}_{N-1} = p'_1, \ldots, p'_{N-1}, \tag{9.6}$$

$$\tilde{q}_1, \ldots, \tilde{q}_{N-1} = q'_1, \ldots, q'_{N-1} \tag{9.7}$$

and

$$p'_N = -\tilde{q}_N \text{ and } q'_N = \tilde{p}_N. \tag{9.8}$$

Equations (9.5)–(9.8) are then used to compute the matrix elements of $\tilde{A}$ in terms of the elements of A' by evaluating partial derivatives of the new Jacobi matrix $D(\tilde{p})/D(\tilde{q})$. The second-derivative coefficients in the new representation are derived by rotating the old coefficients as in Equation (9.3) for consistency, and then substituting the new momenta and coordinates into the old coefficients. For instance, replacing p_N and q_N with $-\tilde{q}_N$ and $\tilde{p}_N$, respectively, gives

$$(H_{\tilde{\mathbf{q}}\tilde{\mathbf{q}}})_{NN} = (H_{\mathbf{pp}})_{NN}. \tag{9.9}$$

Similar trivial substitutions are used to compose all the other elements of the second derivatives in the new representation. The newly composed matrix $\tilde{A}$ also satisfies Equation (9.1) using the modified second derivatives as coefficients. In the new form, the singularity has been eliminated, the propagation runs smoothly through the hitherto divergent region, and the solution of Equation (9.1) can be obtained with the added advantage of locating the caustics by monitoring either its determinants or its eigenvalue. Well beyond the point of divergence, the inverse transformation is carried out in exact reverse order to revert to the matrix A. The original propagation is then continued in regions far away from caustics.

Numerical examples are provided below for the two cases: (1) two-dimensional Henon-Heiles system and (2) three-dimensional adiabatic chemical reaction.

9.1.1 Caustics in Chaotic Henon-Heiles System

The two-dimensional Henon-Heiles Hamiltonian (in atomic units) is given by

$$H = \frac{1}{2}\left(p_x^2 + p_y^2\right) + \frac{1}{2}\left(x^2 + y^2\right) + \left(x^2 y - \frac{1}{3}y^3\right). \tag{9.10}$$

For this Hamiltonian, the Poincaré surface section method is a successful method to distinguish between the regular and chaotic regimes depending on the energy. The destruction of the regular pattern of caustics after the full onset of chaos is a well-known phenomenon; following the work of Stuchi and Vieira-Martins [207], we also seek to locate caustics of these trajectories in both regular and chaotic regimes and compare the picture of caustics in this potential with the surface section at the same initial condition. The classical trajectory was generated from the turning point $[\vec{p}(0) = 0]$, which corresponds to the initial condition $A^{-1} = 0$ for Equation (9.1). After short-time propagation of A^{-1}, the propagation is continued with the matrix A. In the cases considered, the initial x_0 is 0.43 while y_0 has been taken to be -0.37

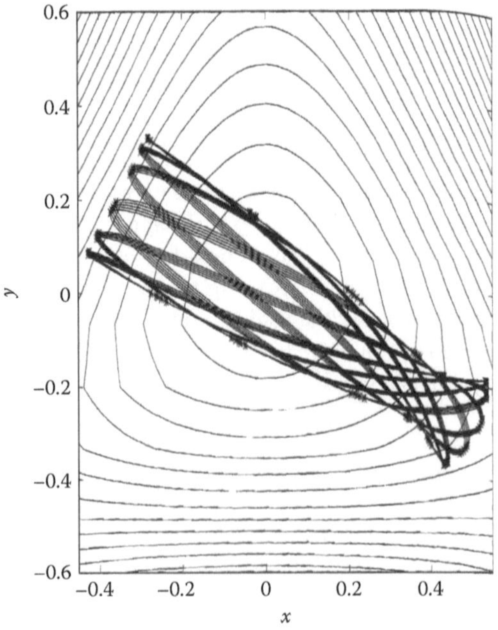

FIGURE 9.1 Trajectory in the Henon-Heiles potential for the initial condition: $x_0 = 0.43$ and $y_0 = -0.37$. The marks (∗) indicate the location of caustics. (Taken from Reference [47] with permission.)

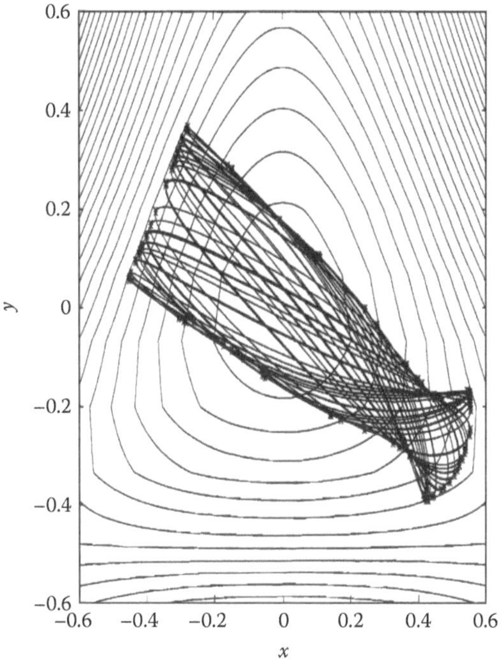

FIGURE 9.2 The same as Figure 9.1 for the initial condition: $x_0 = 0.43$ and $y_0 = -0.39$. Partial destruction of regular caustics is seen. (Taken from Reference [47] with permission.)

(Figure 9.1), −0.39 (Figure 9.2), and −0.41 (Figure 9.4). The location of caustics is marked by $*$. The Poincaré surface sections corresponding to Figure 9.2 and Figure 9.4 are shown in Figure 9.3 and Figure 9.5. These cases from Figure 9.1 to Figure 9.5 correspond to a progressive degeneration of a regular trajectory into full chaos as the accompanying figures show. As is well known, in the case of regular regime (Figure 9.1), the caustics clearly provide the envelope of the family of trajectories. As the system becomes chaotic, discontinuities (see Figure 9.4) and separatrices (see Figure 9.5) appear in the Poincaré surface section and the caustics is no longer discernible with the tori totally destroyed. As can be seen in these figures, the present method works well in all regimes.

9.1.2 Caustics in Chemical Reaction Dynamics

The second example is a three-dimensional triatomic chemical reaction mimicking the reaction $C + H_2 \rightarrow CH + H$ for total angular momentum $J = 0$. The motivation here is naturally to make it possible to take into account the effects of tunneling in chemical reaction dynamics in both classical and semiclassical treatments. In order to run tunneling trajectories in the classically forbidden region, it is inevitable to find caustics along each trajectory in the classically allowed region. We have used the DIM (diatomics-in-molecule) potential matrix of CH_2 [209–211]. The DIM matrix

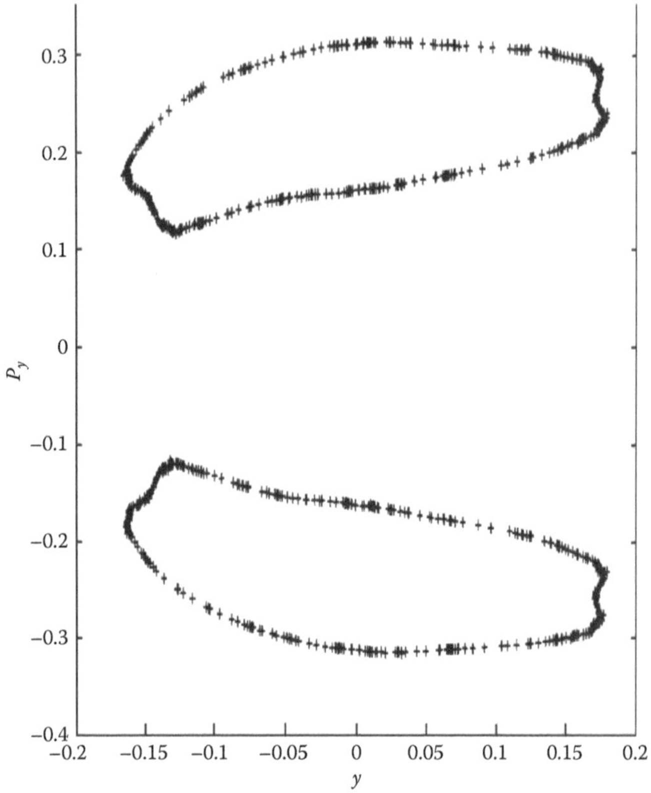

FIGURE 9.3 Poincaré surface corresponding to Figure 9.2. (Taken from Reference [47] with permission.)

is defined as

$$V = \begin{pmatrix} g_1 + \frac{1}{2}(g_3 + h_3 + h_2 + g_2) & \frac{1}{2}(g_3 - h_3) & \frac{1}{2}(g_2 - h_2) \\ \frac{1}{2}(g_3 - h_3) & g_2 + \frac{1}{2}(g_3 + h_3 + h_1 + g_1) & \frac{1}{2}(g_1 - h_1) \\ \frac{1}{2}(g_2 - h_2) & \frac{1}{2}(g_1 - h_1) & g_3 + \frac{1}{2}(g_1 + h_1 + h_2 + g_2) \end{pmatrix}, \tag{9.11}$$

where (h_1, h_2, h_3) and (g_1, g_2, g_3) represent the ground and excited states of H_2, CH_a, and CH_b, respectively. The functional forms of these terms are taken as

$$h_1(r) = -D_h\left[1 + \alpha_1(r - r_{H_2}) + \alpha_2(r - r_{H_2})^2 + \alpha_3(r - r_{H_2})^3\right]\exp\left(-\alpha_4(r - r_{H_2})\right), \tag{9.12}$$

$$g_1(r) = D_g\left[1 + \beta_1 r + \beta_2 r^2\right]\exp(-\beta_3 r), \tag{9.13}$$

$$h_{2,3}(r) = B_h\left(\exp\left[-\gamma_h(r - r_{CH})\right] - 2\right)\exp\left[-\gamma_h(r - r_{CH})\right], \tag{9.14}$$

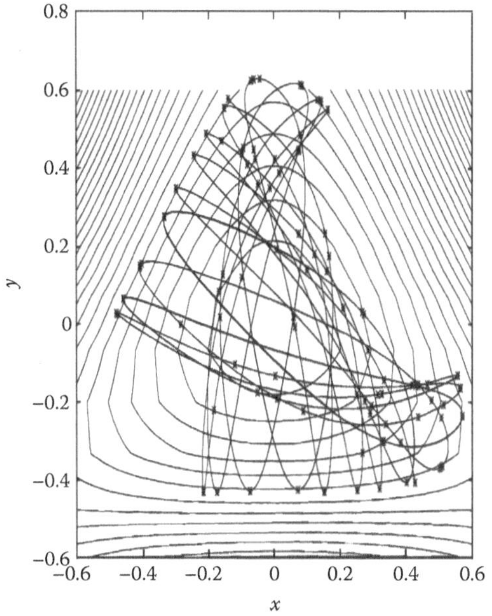

FIGURE 9.4 The same as Figure 9.1 for the initial condition: $x_0 = 0.43$ and $y_0 = -0.41$. Total destruction of regular caustics is seen. (Taken from Reference [47] with permission.)

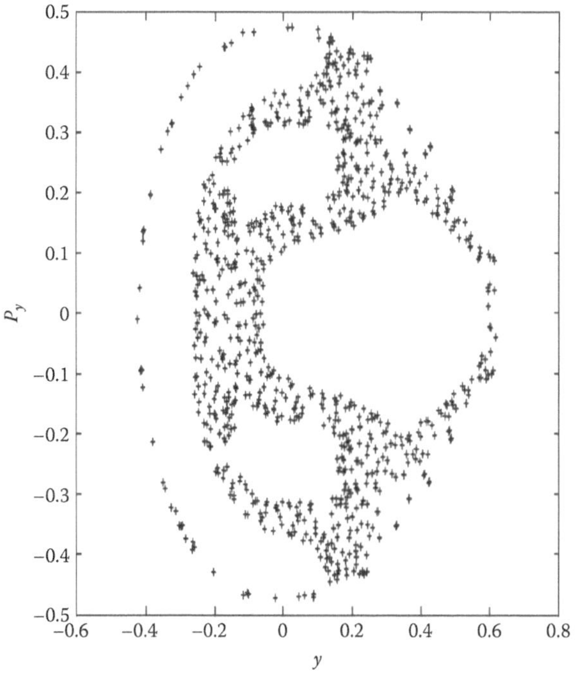

FIGURE 9.5 Poincaré surface corresponding to Figure 9.4. Appearance of separatrices denotes full chaos. (Taken from Reference [47] with permission.)

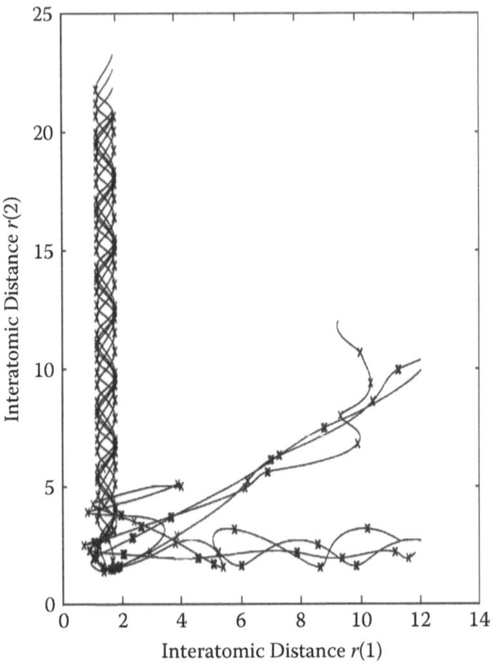

FIGURE 9.6 Family of reactive trajectories in the ground adiabatic potential energy surface determined by Equation (9.11). The crosses indicate caustics. (Taken from Reference [47] with permission.)

TABLE 9.1
Parameters α_i and β_i in Equations (9.12)–(9.13)

i	1	2	3	4
α_i	2.1977034	1.2932502	0.64375666	2.835071
β_i	−1.3874149	0.9098728	2.181301	–

and

$$g_{2,3}(r) = B_g \left(\exp\left[-\gamma_g (r - r_{CH})\right] - 2\right) \exp\left[-\gamma_g (r - r_{CH})\right] . \qquad (9.15)$$

The parameters r_{CH} and r_{H_2} are 2.0 and 1.401 a.u., respectively. The total energy and the initial rovibrational states are taken to be 1.2 eV and ($v = j = 0$). It should be noted that the potential energy surface has an attractive well of depth about 2.3 eV and thus many trajectories are trapped in the well for a long time, as seen in Figure 9.6. Other parameters used are listed in Tables 9.1 and 9.2.

Without loss of generality, a triatomic system can be described by the Jacobi coordinates. Then using the assumption that the center of mass is stationary and

TABLE 9.2
Parameters D_j, B_j, and γ_j in Equations (9.12)–(9.15)

j	h	g
D_j	0.15796326	4.502447
B_j	0.13	0.10
γ_j	1.3	1.5

also the constraint $J = 0$, the system can be described by four coordinates (i.e., the collision is confined to a plane) and thus by a 4×4 matrix. These coordinates are here referred to as X, Y, x, and y. The initial condition for the A-matrix propagation is specified in the asymptotic area where the diatomic molecule is well separated from the colliding partner. Then, the initial conditions for the propagation can be found using well-defined separate integrals of motion in terms of the Jacobi coordinates X, Y, P_X, P_Y and x, y, p_x, p_y for the corresponding arrangement. For instance, for a given initial rovibrational state of the diatomic molecule we have two asymptotic integrals of motion for the internal coordinates,

$$\frac{1}{2m_{H_2}}(p_x^2 + p_y^2) + V(x, y) = \text{constant} \quad \text{and} \quad p_x y - p_y x = \text{constant}, \tag{9.16}$$

and similar relations without potential for the atom-diatom relative coordinates. Differentiating each of these four conservation equations with respect to x, y and X, Y and treating the momenta as functions of the coordinates yield the analytical expression for the matrix $D(P_X, P_Y, p_x, p_y)/D(X, Y, x, y)$ and we obtain the initial condition as

$$\mathbf{A} = \begin{pmatrix} \mathbf{AA} & \mathbf{0} \\ \mathbf{0} & \mathbf{aa} \end{pmatrix}, \tag{9.17}$$

where

$$\mathbf{AA} = \frac{1}{\mathbf{P_X X + P_Y Y}} \begin{pmatrix} P_Y^2 & -P_X P_Y \\ -P_X P_Y & P_X^2 \end{pmatrix} \tag{9.18}$$

and

$$\mathbf{aa} = \frac{1}{\mathbf{p_x x + p_y y}} \begin{pmatrix} -mx\frac{\partial V}{\partial x} + p_y^2 & -mx\frac{\partial V}{\partial y} - p_x p_y \\ -my\frac{\partial V}{\partial x} - p_x p_y & -my\frac{\partial V}{\partial y} + p_x^2 \end{pmatrix}. \tag{9.19}$$

To solve Equation (9.1), we have employed the fourth-order Adams-Bashforth-Moulton predictor-corrector scheme, where the corrector is iterated until the final convergence is achieved. This scheme acts as the first check on the validity of the transformations. Namely, the nonconvergence can flag an error in the solution of Equation (9.1) caused by a defective implementation of the suggested canonical transformations. In order to show the feature of propagation, an example is given in Figure 9.6 that shows a family of reactive trajectories and their numerically determined caustics. As expected, the caustics occur periodically in the asymptotic region corresponding to the turning points of the vibrational motion. The value of the present method is seen in the rearrangement region. Here, the trajectory is no longer periodic and the caustics cannot be easily identified by observation. The method works well to determine those points that are important to incorporate tunneling effects in reaction dynamics.

As mentioned above, it may happen that two or more eigenvalues of matrix A almost simultaneously diverge around the same propagation time, especially when the multidimensional potential has a deep well. This often occurs in the condensation region of the insertion-type chemical reactions. This requires consecutive multiple canonical transformations to treat concomitantly occuring divergences. A bit of care is needed in choosing the time step and divergence criterion. Figure 9.7 shows a time profile of the maximum absolute eigenvalue of the propagated matrix in the condensation region with peaks signifying the occurrence of caustics. In the asymptotic region (not shown here), the time period corresponds to that of vibrational motion of the diatomic molecule. The value of the method of multiple transformations is highlighted by the detection of the internal structure that breaks the periodicity of caustics. Here, the split of the peak corresponds to the situation where three caustics occur within a time interval of about one-tenth of the period of the caustics in the asymptotic region.

The present formulation can be easily applied to any general N-dimensional system. This is a salient point to note, because for any N-dimensional system, the maximum number of divergences that could occur almost in the same time instant is the same as the dimensionality of the system. The method can treat even this extreme case quite capably and is expected to be a potentially powerful complement in the area of classical and semiclassical multidimensional chemical dynamics. As was discussed in Section 3.1, the local separability is assumed at the caustics along each classical trajectory, and the tunneling trajectory is evolved tangentially to the caustic surface on the upside-down potential. Exit to the classically allowed region from this tunneling region can be detected by using the same method discussed above for the upside-down potential. It should be noted that the caustics exists also in the tunneling region between different kinds of C-regions and/or I-regions.

In order to treat electronically nonadiabatic dynamics, the simple TSH (trajectory surface hopping) method has been devised and usefully applied to various processes (see, for instance, [48,113,114,116,117,212,213]). Many classical trajectories representing a given initial quantum state are evolved and the surface hopping probability, or the nonadiabatic transition probability, is calculated for each trajectory when it comes close to the potential energy surface crossing region. If that probability is larger than a random number generated there, the trajectory hops to the other surface. The simplest way of estimating reaction probability is to count the relative number of the trajectories ending up with the required final condition. For the calculations of nonadiabatic transition probability, various methods have been proposed. As was briefly explained in Chapter 5, the analytical Zhu-Nakamura formulas can now be used, enabling us to treat even classically forbidden transitions properly. This method is called the ZN-TSH method and is expected to be useful to attack large chemical and biological systems [48,108,111,112]. The above-mentioned method to treat tunneling process can be incorporated into this ZN-TSH method. Along each tunneling

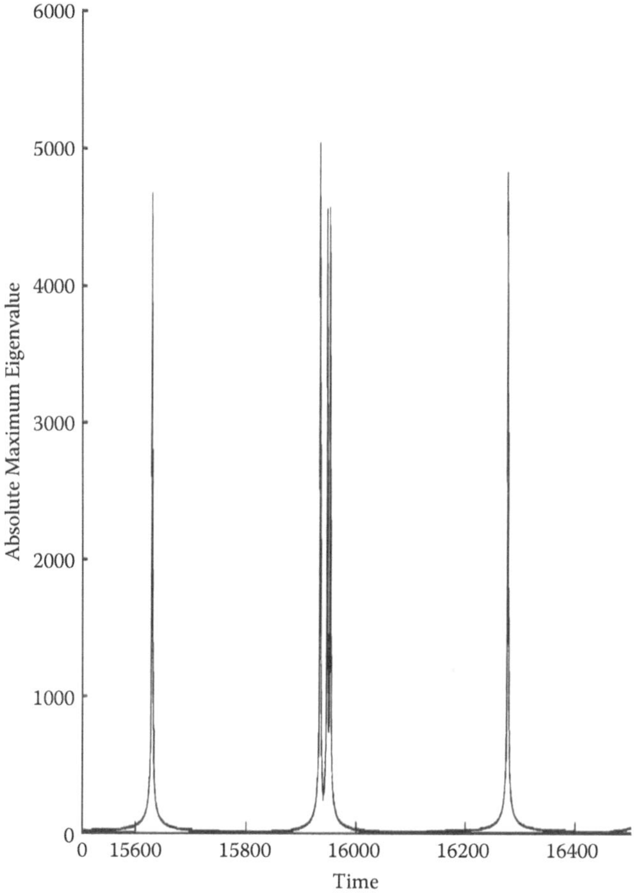

FIGURE 9.7 Time (in a.u.) profile of the absolute maximum eigenvalue of propagated matrix in the case of chemical reaction. Coalescence of caustics occurs in condensation region and appears as split peaks here. (Taken from Reference [47] with permission.)

trajectory the complex action S_{trj} is calculated and the tunneling probability is estimated as $\exp[-2(\mathrm{Im}S_{trj})]/(1+\exp[-2(\mathrm{Im}S_{trj})])$. If the action $|\mathrm{Im}S_{trj}|$ is larger than a given criterion, then that trajectory is not necessary to be further propagated. As a more sophisticated method, the Herman-Kluk semiclassical propagation method can be employed. The initial wave function $\psi_I(\mathbf{r}_0, t = 0)$ is expanded in terms of frozen Gaussian wave packets $g(\mathbf{r}_0; \mathbf{q}_0, \mathbf{p}_0)$ and the latter are propagated along each classical trajectory $(\mathbf{q}_0, \mathbf{p}_0, t = 0) \rightarrow (\mathbf{q}_t, \mathbf{p}_t, t)$. That is to say, the final wave function at time t is expressed as

$$\psi(\mathbf{r}, t) = \int_{traj} \frac{d\mathbf{q}_0 d\mathbf{p}_0}{(2\pi)^N} g(\mathbf{r}; \mathbf{q}_t, \mathbf{p}_t) C_{\mathbf{q}_0, \mathbf{p}_0, t} \exp\left[i S_{\mathbf{q}_0, \mathbf{p}_0, t}\right]$$
$$\times \int d\mathbf{r}_0 g^*(\mathbf{r}_0; \mathbf{q}_0, \mathbf{p}_0) \psi_I(\mathbf{r}_0, t = 0) \tag{9.20}$$

with

$$g(\mathbf{r};\mathbf{q},\mathbf{p}) = \left(\frac{2\gamma}{\pi}\right)^{N/4} \exp[-\gamma(\mathbf{r}-\mathbf{q})^2 + i\mathbf{p}\cdot(\mathbf{r}-\mathbf{q})], \tag{9.21}$$

where γ is a parameter, N is the dimension of the coordinate space, $S_{\mathbf{q}_0,\mathbf{p}_0,t}$ is classical action along the trajectory from $(\mathbf{q}_0, \mathbf{p}_0, t = 0)$ to $(\mathbf{q}_t, \mathbf{p}_t, t)$, and $C_{\mathbf{q}_0,\mathbf{p}_0,t}$ represents the pre-exponential factor obtained from the monodromy matrix [119–121,214]. This is called the Herman-Kluk semiclassical initial value representation (HKSCIVR) and has been usefully applied to multidimensional adiabatic dynamics. As can be easily conjectured, the Zhu-Nakamura nonadiabatic transition amplitude can be incorporated into this formulation so that one can deal with electronically nonadiabatic processes. This is called the ZN-HKSCIVR method and found to be useful to deal with nonadiabatic chemical dynamics [48,108,112,118]. Now, the tunneling processes can also be incorporated into this formulation in the similar way as in the ZN-TSH method mentioned above with the phase contributions taken into account. Incorporation of tunneling effects into HKSCIVR has actually been discussed by many authors [23,215–218]. Kay showed that tunneling can be well incorporated, if a *complex time-dependent* width parameter γ is used [215]. Grossmann proposed a method to treat tunneling by *spawning* classical trajectories [216]. Zhang and Pollak devised a prefactor-free initial value representation and treated tunneling by using segments of classical trajectories by slicing the propagation into intermediate steps [217]. Saltzer and Ankerhold suggested that tunneling is effected by quasi-stationary fluctuation and proposed to use complex trajectories to connect two classical turning points [23,218]. All of these have discussed only the one-dimensional case, but it is now possible to treat multidimensional systems by using the method described in this section.

9.2 DIRECT EVALUATION OF REACTION RATE CONSTANT

9.2.1 Adiabatic Chemical Reaction

Since the thermal reaction rate constant is an important physical quantity to characterize chemical reactions, it is desirable to be able to evaluate the quantity directly without starting from the most detailed quantities of state-to-state reaction probabilities. After the pioneering works of the classical transition state theory (TST) based on the thermal activation done by Eyring and Wigner [219,220], many investigations have been performed to improve this classical TST by including the recrossing and quantum mechanical tunneling effects. See, for instance, the following review articles [11,17,221–223]. The semiclassical instanton approach has also been developed, as explained in Section 3.2 for the case of decay of metastable state. A certain generalization was discussed by Cao and Voth to include the effects of bath modes by using the path integral centroid approach [224]. Recently, the semiclassical instanton approach has been revisited by Kryvohuz and a compact analytical expression of reaction rate constant has been derived [225]. Quantum instanton theory has also been formulated by Miller and co-workers, and various applications to practical systems have been performed [226,227].

Let us introduce here the semiclassical instanton approach to reaction rate constant discussed by Kryvohuz [225]. Using the semiclassical instanton theory based on the

ImF method [17,228–230] and the monodromy matrix [27,231] and introducing an effective one-dimensional model, he derived compact expressions of reaction rate constant [225]. Considering that chemical reaction is a kind of decay or escape process from a metastable state, the rate constant can be calculated by taking a Boltzmann average of the imaginary part of energy of the metastable state. Assuming that the imaginary part of energy is small, the reaction rate constant $k(T)$ at temperature T is explicitly given by

$$k(T) \simeq -\frac{2}{\hbar}\mathrm{Im}F = \frac{2}{\hbar\beta}\mathrm{Im}[\ln(Q)] \simeq \frac{2}{\hbar\beta}\frac{\mathrm{Im}Q}{\mathrm{Re}Q}, \tag{9.22}$$

where

$$\beta = \kappa T, \tag{9.23}$$

F and Q are the Helmholtz free energy and the partition function, respectively, which are defined by

$$Q = \sum_i \exp[-\beta E_i] \equiv \exp(-\beta F). \tag{9.24}$$

As usual, by introducing the imaginary time t by

$$t = -i\hbar\beta \equiv -i\tau_\beta, \tag{9.25}$$

the partition function is expressed in terms of the semiclassical propagator as

$$Q = \mathrm{Tr}(\exp[-\beta H]) = \int d\mathbf{x}\langle\mathbf{x}|\exp[-iHt/\hbar]|\mathbf{x}\rangle = \int D[\mathbf{x}]\exp(i\bar{S}[\mathbf{x}]/\hbar), \tag{9.26}$$

where $D[\mathbf{x}(\tau)]$ represents the path integral and $\bar{S}$ is the action integral of the ordinary Lagrangian with respect to t. By changing the time variable from t to τ, we obtain

$$Q = \oint D[\mathbf{x}(\tau)]\exp(-S[\mathbf{x}(\tau)]/\hbar), \tag{9.27}$$

where $\oint \equiv \int d\mathbf{x}_i \int_{\mathbf{x}(0)=\mathbf{x}_i}^{\mathbf{x}(\hbar\beta)=\mathbf{x}_i}$ and $S[\mathbf{x}(\tau)]$ is the action integral of the Lagrangian with upside-down potential with respect to τ and is given by

$$S[\mathbf{x}(\tau)] = \int_0^{\tau_\beta}\left[\sum_{n=1}^{N}\frac{m}{2}\left(\frac{dx_n}{d\tau}\right)^2 + V[\mathbf{x}(\tau)]\right]d\tau, \tag{9.28}$$

where m is the mass and N represents the dimension of coordinate space $\mathbf{x}$. As was mentioned before in Section 3.2 [below Equation (3.30)], taking the integration over the range $(0, \tau_\beta)$ provides the rate at a given temperature. There are two solutions of the stationarity condition of the action, $\delta S = 0$. One is the trajectory sitting on the maximum of the inverted potential that gives the partition function Q_r of the reactant, which is equal to ReQ. The second solution provides the classical instanton trajectory $\mathbf{x}_{inst}(\tau)$ and the corresponding partition function becomes imaginary. Thus we have

$$k(T) = \frac{2}{\hbar\beta Q_r}\mathrm{Im}[Q(\mathbf{x}_{inst})], \tag{9.29}$$

where $\mathbf{x}_{inst}$ represents the instanton trajectory in the N-dimensional coordinate space. To evaluate $Q(\mathbf{x}_{inst})$ we introduce one-dimensional coordinate X along the classical instanton trajectory $\mathbf{x}_{inst}(\tau)$ and an $(N-1)$-dimensional vector $\mathbf{Y}$ for the transverse displacements from the instanton trajectory. Assuming that the transverse quantum fluctuations $\delta\mathbf{Y}$ are uncoupled from the longitudinal fluctuation δX, we can expand the action in Equation (9.28) up to quadratic terms in $\delta\mathbf{Y}$ [17,230,232,233] as

$$S \simeq \int_0^{\tau_\beta} \left[\frac{m}{2} \left(\frac{dX}{d\tau} \right)^2 + V[X(\tau)] \right] d\tau + \sum_{i,j} \frac{1}{2} \int_0^{\tau_\beta} \left[\delta_{ij} m \left(\frac{dY_i}{d\tau} \right)^2 + V''_{ij}[X(\tau)] Y_i Y_j \right] d\tau, \tag{9.30}$$

where $X(\tau) = X_{inst}(\tau) + \delta X$ and $\mathbf{Y} = \delta\mathbf{Y}$. The second derivative $V''_{ij}[X(\tau)]$ is taken at $\{X = X(\tau), \mathbf{Y} = 0\}$ and parametrically depends on $X(\tau)$. The integration over $\mathbf{Y}$ degrees of freedom can be carried out exactly and gives [230,232]

$$\prod_{n=1}^{N-1} \frac{1}{2\sinh(\lambda_n[X(\tau)]/2)}, \tag{9.31}$$

where $\exp(\pm\lambda_n)(n = 1 \sim N-1)$ are eigenvalues of the $2N \times 2N$ monodromy matrix $\mathbf{R}_{2N}(\tau_\beta)$, which satisfies the following differential equation along the instanton trajectory:

$$\frac{d}{d\tau}\mathbf{R}_{2N}(\tau) + \mathbf{F}(\tau)\mathbf{R}_{2N}(\tau) = 0 \tag{9.32}$$

with

$$\mathbf{F}(\tau) = \begin{pmatrix} \mathbf{0} & -\mathbf{1/m} \\ \frac{\partial^2 V}{\partial x_n \partial x_m} & \mathbf{0} \end{pmatrix}_{inst} \tag{9.33}$$

and

$$\mathbf{R}_{2N}(\tau = 0) = \mathbf{1}. \tag{9.34}$$

In the case of harmonic potential, the stability parameters λ_n are explicitly given by

$$\lambda_n = \hbar\beta\omega_n, \tag{9.35}$$

where ω_n are the harmonic frequencies of transverse degrees of freedom along the instanton trajectory. $\mathbf{R}_{2N}$ has two more eigenvalues corresponding to the zero stability parameter $\lambda = 0$ that correspond to the instanton degree of freedom $X(\tau)$ and manifests its invariance with respect to time shift. Now, the rate constant $k(T)$ is expressed as [230]

$$k(T) = \frac{2}{\hbar\beta Q_r} \mathrm{Im} \oint D[X(\tau)] \exp\left[-S_{eff}[X(\tau)]/\hbar\right], \tag{9.36}$$

where

$$S_{eff} = S_0[X(\tau)] + \sigma[X(\tau)] \equiv \int_0^{\tau_\beta} \left[\frac{m}{2} \left(\frac{dX}{d\tau} \right)^2 + V[X(\tau)] \right] d\tau + \sigma[X(\tau)] \tag{9.37}$$

with

$$\sigma[X(\tau)] = \hbar \sum_{n=1}^{N-1} \ln[2\sinh(\lambda_n[X(\tau)]/2]. \tag{9.38}$$

Since the stability parameters λ_n depend on the quantum trajectory $X(\tau)$, the effective classical trajectory of instanton is the solution of $\delta S_{eff}/\delta X = 0$, namely,

$$\int_0^{\tau_\beta} \left[-m\frac{\partial^2 X}{\partial \tau^2} + \frac{\partial V(\tau)}{\partial X} \right] d\tau + \frac{\hbar}{2} \sum_n \frac{\delta \lambda_n}{\delta X} \coth(\lambda_n/2) = 0, \tag{9.39}$$

where the last term introduces a kind of additional potential. Once the effective instanton $X(\tau)$ is solved, the overall effective one-dimensional potential is given by

$$V_{eff}(X) = V(X) + \frac{1}{2}\sum_n \hbar\omega_n[X(\tau)], \tag{9.40}$$

where $\omega_n[X(\tau)]$ are harmonic frequencies of the transverse degrees of freedom along the instanton trajectory. The instanton energy $E(\beta)$ is obtained from

$$E(\beta) = \frac{\partial S_{eff}}{\partial \tau}|_{\tau=\tau_\beta}. \tag{9.41}$$

This formulation corresponds to the improved second approximation in Reference [225]. In the simpler first approximation [225], the instanton is determined by the first term of Equation (9.39) (Gutzwiller approximation), namely by

$$-m\frac{\partial^2 X}{\partial \tau^2} + \frac{\partial V(\tau)}{\partial X} = 0. \tag{9.42}$$

In this approximation λ_n depends only on the instanton determined by this condition and we simply have

$$\lambda_n[X(\tau)] = \lambda_n[X_{inst}(\tau)] = \lambda_n(\tau_\beta), \tag{9.43}$$

$$\sigma[X(\tau_\beta)] = \sigma(\tau_\beta), \tag{9.44}$$

and

$$S_{eff} \simeq S_0(\tau_\beta) + \sigma(\tau_\beta). \tag{9.45}$$

Clearly, this first approximation is much simpler than the second one mentioned above. Both first and second approximations are applied to several collinear chemical reactions [225]. The first approximation was found to work well for the simplicity compared with the second one. Once the instanton trajectory $X_{inst}(\tau)$ and the effective one-dimensional potential $V_{eff}(X_{inst})$ are determined, the rate constant is calculated from the one-dimensional transmission corresponding to S_{eff}. In the case of the first approximation, we have

$$k(T) = \frac{Q_{trans}}{Q_r} f, \tag{9.46}$$

where the flux f is given by

$$f = \frac{1}{2\pi\hbar}\int_{-\infty}^{V_0} e^{-\beta E} e^{-W(E)} dE \tag{9.47}$$

with

$$W(E) = S_0(\tau_\beta) - \tau_\beta \frac{\partial S_{eff}}{\partial \tau_\beta} = S_0(\beta) - \tau_\beta E(\beta), \tag{9.48}$$

where V_0 is the barrier height of the one-dimensional effective potential. Equation (9.47) is valid for the temperature lower than the upper limit at which the instanton trajectory exists. Expanding $W(E)$ around $E = E_0(\beta)[= (\partial S_0/\partial\beta)/\hbar]$ up to the second order and performing the Gaussian-type integration in Equation (9.47) with use of the relation

$$\frac{\partial^2 W}{\partial E^2}|_{E_0(\beta)} = -\left[\frac{\partial E}{\partial \beta}\right]^{-1}|_{E_0(\beta)}, \tag{9.49}$$

we finally obtain the compact expression of the rate constant as

$$k = \frac{Q_{trans}}{Q_r}\sqrt{\frac{-E_0'(\beta) - \sigma''(\beta)/\hbar}{2\pi\hbar^2}}\exp[-S_0(\beta)/\hbar]\mathrm{erf}\left[\frac{V_0 - E_0(\beta) - \sigma'(\beta)/\hbar}{\sqrt{-E_0'(\beta) - \sigma''(\beta)/\hbar}}\right], \tag{9.50}$$

where

$$\mathrm{erf}(x) = \frac{1}{\sqrt{2\pi}}\int_{-\infty}^{x} \exp(-t^2/2)dt \tag{9.51}$$

and

$$Q_{trans} = \prod_{n=1}^{N-1} \frac{1}{2\sinh\left(\frac{1}{2}\lambda_n(\tau_\beta)\right)}. \tag{9.52}$$

Further detailed discussions and applications to collinear chemical reactions are given in Reference [225]. Miller and co-workers have formulated the *quantum* instanton theory, starting from the expression of rate constant in terms of the flux operators, and applied it to various practical processes. Those who are interested in that should refer to References [226,227]. The semiclassical instanton theory explained above can be conveniently generalized to the case of nonadiabatic chemical reaction. This is discussed in the next subsection.

9.2.2 Nonadiabatic Chemical Reaction

As for electronically nonadiabatic chemical reactions, one typical example of which is the electron transfer process between molecules [234–236], not many works have been done so far to formulate a general theory for directly evaluating the reaction rate constant beyond the perturbative treatment of electronic coupling. Voth and co-workers [237] have developed a computational methodology to evaluate rate constant including the effects of bath by using the Pechukas-type treatment of nonadiabatic transition based on the diabatic state representation [238]. Okuno and Mashiko formulated an expression of rate constant by using the Landau-Zener formula at the minimum energy crossing point [239]. A nonadiabatic version of transition state theory for unimolecular reactions was developed by Marks and Thompson by using the phase space ensemble and the Landau-Zener formula in the diabatic state representation [240]. This was applied to bond fission processes in bromoacetyl chloride [241]. Topaler and Truhlar studied nonadiabatic decay rate of excited complex in scope of statistical model [242]. Nakamura and co-workers [243] formulated a nonadiabatic transition state theory by using the Miller's general formula of the flux-side correlation function [244] and the Zhu-Nakamura theory of nonadiabatic transition [48]. They have applied it to electron transfer and are successful to improve the celebrated Marcus formula by uniformly covering the whole range of electronic coupling strength in the harmonic oscillator approximation [245]. There have been proposed various simulation methods to deal with nonadiabatic processes (see, for instance, [48,108]). The simplest ones are the trajectory surface hopping (TSH) methods such as the fewest switches TSH [113,246] and the ZN-TSH method mentioned in Sections 5.2 and 9.1 [111,112]. A bit more sophisticated ones to include the effects of phases are the ZN-HKSCIVR method, also mentioned in these sections [118], and the spawning method developed by Levine and Martinez [247]. It should be noted that these methods are not for evaluating the rate constants directly. It should also be noted that only the Zhu-Nakamura formulas can treat classically forbidden transitions properly and analytically.

In this subsection a formulation of compact expression for the nonadiabatic reaction rate constant is proposed [248]. Since the nonadiabatic chemical dynamics would open up a new significant branch of chemical dynamics not only in the sense of comprehending the dynamics occuring in nature but also in the sense of controlling the dynamics and manifesting new molecular functions [112], a compact formulation of the nonadiabatic chemical reaction rate constant would be quite useful. The semiclassical instanton approach of Kryvohuz introduced in the previous subsection [225] may be generalized to nonadiabatic processes. The instanton trajectory is defined by including the effects of nonadiabatic transition and searched by employing the method similar to that used in tunneling splitting. Furthermore, the analytical Zhu-Nakamura formulas of nonadiabatic transitions can be incorporated. This theory can cover a wide energy range from deep tunneling to high energy above the crossing point. For clarity, we consider the nonadiabatic tunneling-type curve crossing shown in Figure 5.4, although it is possible to treat more general cases. As explained in Chapter 5, it is important to note that the tunneling through the lower adiabatic potential $E_1(X)$ is affected by the upper adiabatic potential $E_2(X)$. The effective action, S_{eff}, of Equation (9.37) in the first approximation should now be replaced by

$$S_{eff}^{ND}[X(\tau_\beta)] = \int_0^{\tau_\beta} \left[\frac{m}{2} \left(\frac{dX}{d\tau} \right)^2 + V[X(\tau)] \right] d\tau$$
$$+\sigma(\tau_\beta) - \ln(1 - p_{ZN}) + 2\hbar\phi_S[X(\tau_\beta)] \tag{9.53}$$

where $\sigma(\tau_\beta)$ is the contribution from the transverse degrees of freedom as in the adiabatic case and ϕ_S is the so-called dynamical phase due to nonadiabatic transition that is given by Equation (5.49) in Section 5.2. The potential $V(X)$ represents the upside-down of $E_1(X)$. When the instanton trajectory runs through the *upside-down* adiabatic potential $-E_1(X)$, the phase ϕ_S and the factor $\ln(1 - p_{ZN})$, which represents the transferability, should be added. Actually, the transition amplitude I_{11} for staying on the potential $-E_1(X)$ by one passage of the avoided crossing point is given by

$$I_{11} = \sqrt{1 - p_{ZN}} \exp(i\phi_S), \tag{9.54}$$

where p_{ZN} is the Zhu-Nakamura nonadiabatic transition probability given by Equation (5.45). The instanton trajectory is determined from the stationarity condition as before,

$$\frac{\delta S_{eff}^{ND}}{\delta X} = 0. \tag{9.55}$$

When the instanton trajectory $X_{inst}(\tau)(\tau = 0 \to \tau = \tau_\beta)$ is determined from this condition, the one-dimensional effective potential $V_{eff}(X_{inst})$ is obtained for the given temperature β. Thus the original multidimensional problem is reduced to the one-dimensional transmission through this effective potential along the instanton trajectory and the rate constant $k(T)$ is given by

$$k(T) = \frac{Q_{trans}}{Q_r} f_{ND}, \tag{9.56}$$

where Q_{trans} is the contribution from the transversal degrees of freedom and is given by Equation (9.52). The rate f_{ND} is the nonadiabatic transmission rate given by

$$f_{ND} = \frac{1}{2\pi\hbar}\int_{-\infty}^{E_t} dE \exp(-\beta E)P(E), \tag{9.57}$$

where E_t is the barrier top of the lower adiabatic potential (see Figure 5.4) and $P(E)$ is the nonadiabatic tunneling probability given by the Zhu-Nakamura theory. Since we are considering the energy region $E \le E_t$ here, the transmission probability $P(E)$ is given by Equation (5.24). However, since the transmission probability expressions are also available at higher energies $E \ge E_t$, the rate constant at these energies can be given by the same expression as Equation (9.56). Since the instanton trajectory shrinks to a point X_t, top of the potential barrier, we have

$$Q_{trans} = \prod_{n=1}^{N-1} \frac{1}{2\sinh\left(\frac{\tau_\beta}{2}\omega_n^\dagger\right)}, \tag{9.58}$$

where $\omega_n^\dagger$ are transverse frequencies at X_t. The flux f_{ND} is the same as Equation (9.57) with $P(E)$ given by Equations (5.37) and (5.46) for energies $E_t \le E \le E_b$ and $E \ge E_b$, respectively.

The instanton trajectory can be searched for by using a method similar to that used in Section 6.2. The instanton trajectory is a periodic orbit starting from $\mathbf{x}_i$ at $\tau = 0$ and coming back to the same position $\mathbf{x}_i$ at $\tau = \tau_\beta$. The starting point $\mathbf{x}_i$ corresponds to the turning point T_1^l in Figure 5.4 at energy $E(\beta)$. At the half time $\tau = \tau_\beta/2$ it reaches the other turning point $\mathbf{x}_f$, which corresponds to T_1^r in Figure 5.4. Instead of using the time τ, we introduce the following parameter z:

$$z = \frac{1}{\tau_\beta}(4\tau - \tau_\beta), \tag{9.59}$$

where $\tau = 0(\mathbf{x} = \mathbf{x}_i)[\tau = \tau_\beta/2(\mathbf{x} = \mathbf{x}_f)]$ corresponds to $z = -1[z = 1]$. The instanton trajectory becomes a parametrized path and is expressed as

$$x_n = \frac{1-z}{2}x_{in} + \frac{1+z}{2}x_{fn} + (1-z^2)\sum_{l=1}^{M} C_{ln}\phi_l(z), \tag{9.60}$$

where $\phi_l(z)$ are taken to be Legendre polynomials. The action S_{eff}^{ND} is expressed as a function of $\{\mathbf{C}_l, \mathbf{x}_\alpha(\alpha = i, f)\}$. It should be noted that being different from the tunneling splitting case, the initial and final positions $\mathbf{x}_\alpha(\alpha = i, f)$ are unknown and should be determined variationally. Since there is a symmetry for $\tau = 0 \rightarrow \tau_\beta/2$ and $\tau = \tau_\beta/2 \rightarrow \tau_\beta$, what we actually need is the half-period calculation from $\tau = 0$ to $\tau = \tau_\beta/2$. The $\tau \rightarrow z$ transformation in Equation (9.59) is a simple linear one, being quite different from the nonlinear one used in Section 6.2. Because of this and the necessity to determine the endpoints $\mathbf{x}_\alpha(\alpha = i, f)$, the convergence is much slower with respect to M. The procedure to find the instanton path for each temperature β is as follows:

1. Assuming certain initial values for ϕ_D, $\{\mathbf{C}_l\}(l = 1, \ldots, M)$ and $\mathbf{x}_\alpha(\alpha = i, f)$, we calculate approximate path, instanton energy, and avoided crossing point $\mathbf{x}_c$. Then we can compute $\sigma(\tau_\beta)$ and $\phi_D[X(\tau)]$, and determine the zeroth order approximate instanton path.
2. Using the previous results of $\{\mathbf{C}_l, \mathbf{x}_\alpha\}$, we apply the following variational principle. Namely, for the variation

$$\{\mathbf{C}_l\} \rightarrow \{\mathbf{C}_l + \delta\mathbf{C}_l\} \text{ and } \mathbf{x}_\alpha \rightarrow \mathbf{x}_\alpha + \delta\mathbf{x}_\alpha(\alpha = i, f), \tag{9.61}$$

we minimize the action

$$\frac{\partial \delta S_{eff}^{ND}(\{\delta\mathbf{C}_l\}, \delta\mathbf{x}_\alpha)}{\partial \delta\mathbf{C}_l} = 0 \tag{9.62}$$

and

$$\frac{\partial \delta S_{eff}^{ND}(\{\delta\mathbf{C}_l\}, \delta\mathbf{x}_\alpha)}{\partial \delta\mathbf{x}_\alpha} = 0. \tag{9.63}$$

From this improved path, $\sigma(\tau_\beta)$ is calculated and the improved $\phi_D[X(\tau)]$ is obtained. It should also be confirmed whether $E(\beta) = E_1(\mathbf{x}_i) = E_1(\mathbf{x}_f)$ is satisfied. Then the improved instanton path is determined and we repeat this procedure.
3. If everything is converged, we go back to the process (2).

10 Concluding Remarks and Future Perspectives

Such a basic idea of the instanton theory that the energy E is treated as the first-order term of $\hbar$ is useful for the problems of tunneling splitting in symmetric double well potential and decay of metastable state through tunneling. With use of this basic idea, the modified WKB theory was formulated and demonstrated to be applicable virtually to any real molecular systems with high accuracy. There are two key points that made this possible. The first is that we can find the instanton trajectory relatively easily and accurately in multidimensional space. The second is that the high level of *ab initio* quantum chemical data along the instanton trajectory can be efficiently used to calculate desirable physical quantities with no need to construct a global potential energy surface in multidimensional space. Although the high-level quantum chemical calculations—including first- and second-order derivatives of potential energies—require a bit too much CPU time, this can be alleviated by using lower-level non-time-consuming quantum chemical calculations in an appropriate way. This recipe was explained in detail. Numerical applications to real polyatomic molecules have been carried out and demonstrated to reproduce the highly accurate spectroscopic experimental data. Furthermore, insufficiency of low (one and two) dimensional treatments and the effects of multidimensionality were clearly pointed out and numerically demonstrated. Vibrational excitations do not necessarily enhance tunneling, and tunneling probabilities may decrease or oscillate with the excitation, depending on the topology of potential energy surface. Although full-dimensional treatment is possible, from the viewpoint of comprehending the tunneling dynamics and avoiding unnecessary heavy computations, it is desirable to find out which degrees of freedom are crucial to be included in order to comprehend the tunneling dynamics properly. An example was provided in Section 7.4 in which the out-of-plane wagging motion is found to play important roles in the case of malonaldehyde. Analyses along this line should be further carried out. It is also necessary to develop theories further so that we can deal with rovibrationally excited states in general. Recently, the ring-polymer instanton method has been developed by Althorpe and co-workers for calculating tunneling splitting in multidimensional systems [249]. Some detailed studies would be necessary to compare with other methods such as the one presented here.

Although mathematical development is still required to improve the semiclassical mechanics to cover both classically allowed and forbidden regions, multidimensional tunneling phenomena in chemical reactions or more generally in various chemical dynamics can now be dealt with to a good extent by using classical trajectories. In classically allowed regions, ordinary classical trajectories can be run and the caustics that defines the boundary between classically allowed and forbidden regions can be determined relatively easily. This detection of caustics in high-dimensional space cannot be done by simply running classical trajectories, but can be made by solving an appropriate stable Riccati-type nonlinear differential equation along a classical trajectory. Assuming the local separability at caustics, we can run classical trajectories on the inverted upside-down potential in the classically forbidden region. This kind of formulation can be incorporated into the classical and semiclassical methods that can deal with nonadiabatic dynamics such as ZN-TSH and ZN-HKSCIVR methods based on the Zhu-Nakamura theory of nonadiabatic transition. Then, such important quantum mechanical effects as nonadiabatic transition, tunneling, and various phases can be taken into account by running classical trajectories. These methods can be employed within the on-the-fly framework and are expected to be applied to a wide range of chemical and biological dynamic processes [22]. Tunneling phenomena would be important not only at low energies on the electronically ground potential energy surface, but also at high energies on the electronically excited potential energy surface, whenever potential barriers exist. Theories for direct evaluation of reaction rate constants have also been well developed for electronically adiabatic chemical reactions such as the variational transition state theory, and semiclassical and quantum instanton theories. The semiclassical instanton approach for nonadiabatic chemical reactions was also presented by generalizing the corresponding approach for adiabatic chemical reactions. This kind of treatment has to be further improved, such as in the treatment of Chapter 6. In chemical and biological dynamics, nonadiabatic tunneling-type transitions are rather ubiquitous and care should always be taken whether the potential barrier encountered is an ordinary adiabatic potential barrier or created by the interaction with higher excited state, namely, the nonadiabatic tunneling-type barrier. One-dimensional theory of the nonadiabatic tunneling phenomenon has been established by Zhu and Nakamura, but multidimensional theory has yet to be developed.

There are many other important topics that could not be treated in this book such as effects of environment, chaos-assisted tunneling or chaotic tunneling, time-dependent tunneling phenomena, and macroscopic tunneling. Those who are interested in these topics should refer to the following books [17,21–24,233,250,251]. Tunneling time, which cannot be defined definitely, presents another interesting subject that we did not discuss in this book. The reader should refer to the books [21,48], for instance.

A Proofs of Equation (2.95) and Equation (2.110)

By following Appendix A of [2] by Coleman, Equation (2.110) and Equation (2.95) are proved. Let us consider the equation

$$\left(-\frac{\partial^2}{\partial \tau^2} + W\right) \psi_\lambda(\tau) = \lambda \psi_\lambda(\tau), \tag{A.1}$$

where W is some bounded function of τ. The solution function that satisfies the boundary conditions,

$$\psi_\lambda(-T/2) = 0 \text{ and } \frac{\partial \psi_\lambda}{\partial \tau}(\tau = -T/2) = 1, \tag{A.2}$$

is denoted as $\psi_\lambda(\tau)$. The operator $-\partial^2/\partial\tau^2 + W$ has an eigenvalue λ_n, if and only if the eigenfunction $\psi_{\lambda_n}(\tau)$ satisfies the boundary condition

$$\psi_{\lambda_n}(\tau = T/2) = 0. \tag{A.3}$$

Let $W^{(1)}$ and $W^{(2)}$ be two functions of τ and the corresponding solutions of Equation (A.1) are denoted as $\psi_\lambda^{(1,2)}(\tau)$. We prove the following equation:

$$\det\left[\frac{-\frac{\partial^2}{\partial\tau^2} + W^{(1)} - \lambda}{-\frac{\partial^2}{\partial\tau^2} + W^{(2)} - \lambda}\right] = \frac{\psi_\lambda^{(1)}(T/2)}{\psi_\lambda^{(2)}(T/2)}. \tag{A.4}$$

The left-hand side of this equation is a meromorphic function of λ with a simple zero at each $\lambda_n^{(1)}$ and a simple pole at $\lambda_n^{(2)}$. By the Fredholm theorem of complex function theory, the left-hand side goes to one as λ goes to infinity in any direction except along the positive real axis. The right-hand side is also a meromorphic function with exactly the same zeros and poles. By elementary differential equation theory, it also goes to one in the same limit. Thus the ratio of the two sides is an analytic function of λ that goes to one as λ goes to infinity in any direction except along the positive real axis. That is to say, the ratio is equal to one. This proves the above equality.

Noting that Equation (2.108) has the primed $\det'$ that excludes the lowest eigenvalue, we can easily see that Equation (2.109) is equal to Equation (2.110).

If we define a normalization quantity N by

$$\frac{\det\left(-\frac{\partial^2}{\partial\tau^2} + W\right)}{\psi_{\lambda=0}(T/2)} = \pi\hbar N^2, \tag{A.5}$$

then by Equation (A.4) N is independent of W. Thus we have the following relation:

$$N\left[\det\left(-\frac{\partial^2}{\partial\tau^2} + W\right)\right]^{-1/2} = [\pi\hbar\psi_0(T/2)]^{-1/2}. \tag{A.6}$$

As a simple example, let us consider the harmonic oscillator $W = \omega^2$. We have the solution of Equation (A.1) under the boundary conditions Equation (A.2) as

$$\psi_0(\tau) = \omega^{-1} \sinh[\omega(\tau + T/2)]. \tag{A.7}$$

Since this solution goes to $(1/2\omega)\exp[\omega T]$ at $T \to \infty$, we can easily obtain Equation (2.95).

B Derivation of Equation (6.80)

Noting Equations (6.17) and (6.9), let us consider the integral

$$\int d\mathbf{q}\delta\left(\sum_i C_i q_i\right)\exp\left(-A_{ij}q_iq_j\right) \equiv \int d\mathbf{q}\delta\left(\mathbf{C}^T\mathbf{q}\right)\exp\left(-\mathbf{q}^T\mathbf{A}\mathbf{q}\right). \tag{B.1}$$

Making the orthogonal transformation $\mathbf{q} = \mathbf{T}\mathbf{u}$, we rewrite this integral as

$$\int d\mathbf{u}\delta\left(\vartheta^T\mathbf{u}\right)\exp\left(-\mathbf{u}^T\Lambda\mathbf{u}\right), \tag{B.2}$$

where

$$\vartheta = \mathbf{T}^T\mathbf{C} \tag{B.3}$$

and

$$\Lambda = \mathbf{T}^T\mathbf{A}\mathbf{T}. \tag{B.4}$$

We can always choose the orthogonal matrix $\mathbf{T}$ in such a way that ϑ has only one nonzero component

$$\vartheta_i = \delta_{iN}\sqrt{\mathbf{C}^T\mathbf{C}}. \tag{B.5}$$

In this case the integration gives

$$\int d\mathbf{u}\delta(\vartheta_N u_N)\exp\left(-\mathbf{u}^T\Lambda\mathbf{u}\right) = \frac{1}{\vartheta_N}\sqrt{\frac{\pi^{N-1}}{\det\tilde{\Lambda}}}, \tag{B.6}$$

where $\tilde{\Lambda}$ is the $(N-1)$ by $(N-1)$ matrix obtained from Λ by deleting the Nth column and Nth row. Taking into account the relation between the minors of the matrix and its inverse, we get

$$\frac{1}{\vartheta_N}\sqrt{\frac{\pi^{N-1}}{\det\tilde{\Lambda}}} = \sqrt{\frac{\pi^{N-1}}{\det\Lambda(\vartheta^T\Lambda^{-1}\vartheta)}} = \sqrt{\frac{\pi^{N-1}}{\det\mathbf{A}(\mathbf{C}^T\mathbf{A}^{-1}\mathbf{C})}}. \tag{B.7}$$

This proves the third equal of Equation (6.80). Referring to the discussion in Section 6.1.1 around Equation (6.9) and noting

$$\tau_{iN} = \frac{dq_{0i}}{dS} = \frac{dq_{0i}}{d\tau}/\frac{dS}{d\tau} = \frac{p_i}{|p|}, \tag{B.8}$$

we can obtain the final expression of Equation (6.80).

C Herring Formula in Curved Space

The derivation [see Equation (6.84)] is quite similar to the Cartesian case. We consider two Schrödinger equations, one for state n,

$$\left[-\frac{1}{2\sqrt{g}}\frac{\partial}{\partial q^i}\sqrt{g}g^{ij}\frac{\partial}{\partial q^j} + V(q)\right]\Psi_n(q) = E_n\Psi_n(q), \tag{C.1}$$

and the one for state m. Next we multiply the first equation by Ψ_m, the second by Ψ_n, subtract one from the other, and integrate with the measure $\sqrt{g}$ over the part of the coordinate space V_Σ with the boundary Σ of the dividing surface. We get

$$\Delta E\int_{V_\Sigma}\sqrt{g}\Psi_n\Psi_m dq = \frac{1}{2}\sum_{ij}\int_{V_\Sigma} d\mathbf{q}\left[\Psi_n\frac{\partial}{\partial q^i}\sqrt{g}g^{ij}\frac{\partial}{\partial q^j}\Psi_m - \Psi_m\frac{\partial}{\partial q^i}\sqrt{g}g^{ij}\frac{\partial}{\partial q^j}\Psi_n\right]$$

$$\times\frac{1}{2}\sum_{ij}\int dq_1\cdots dq_{i-1}dq_{i+1}\cdots dq_N\int^{q_i(q_1\cdots q_{i-1},q_{i+1}\cdots q_N)} dq_i$$

$$\times\left[\Psi_n\frac{\partial}{\partial q^i}\sqrt{g}g^{ij}\frac{\partial}{\partial q^j}\Psi_m - \Psi_m\frac{\partial}{\partial q^i}\sqrt{g}g^{ij}\frac{\partial}{\partial q^j}\Psi_n\right], \tag{C.2}$$

where $q_i(q_1\cdots q_{i-1}, q_{i+1}\cdots q_N)$ is the solution of the equation $f(q) = 0$, which determines Σ. For any i the integration by parts gives

$$\sum_j\int^{q_i(q_1\cdots q_{i-1},q_{i+1}\cdots q_N)} dq_i\left[\Psi_n\frac{\partial}{\partial q^i}\sqrt{g}g^{ij}\frac{\partial}{\partial q^j}\Psi_m - \Psi_m\frac{\partial}{\partial q^i}\sqrt{g}g^{ij}\frac{\partial}{\partial q^j}\Psi_n\right]$$

$$= \sum_j\sqrt{g}\left[\Psi_n\frac{\partial}{\partial q^i}\sqrt{g}g^{ij}\frac{\partial}{\partial q^j}\Psi_m - \Psi_m\frac{\partial}{\partial q^i}\sqrt{g}g^{ij}\frac{\partial}{\partial q^j}\Psi_n\right]\Big|_{q_i=q_i(q_1\cdots q_{i-1},q_{i+1}\cdots q_N)}$$

$$= \sum_j\int dq_i\sqrt{g}\delta(f(q))\frac{\partial f}{\partial q^i}\left[\Psi_n g^{ij}\frac{\partial\Psi_m}{\partial q^j} - \Psi_m g^{ij}\frac{\partial\Psi_n}{\partial q^j}\right] \tag{C.3}$$

and Equation (C.2) can be rewritten as

$$\Delta E\int_{V_\Sigma} dq\sqrt{g}\Psi_n\Psi_m = \frac{1}{2}\sum_{ij}\int dq\sqrt{g}\delta\left(f(q)\right)\frac{\partial f}{\partial q^i}\left[\Psi_n g^{ij}\frac{\partial\Psi_m}{\partial q^j} - \Psi_m g^{ij}\frac{\partial\Psi_n}{\partial q^j}\right]. \tag{C.4}$$

To calculate the tunneling splitting we take

$$\Psi_n(q) = \Psi_{sym}(q) \equiv \frac{\Psi(q) + \Psi(Cq)}{\sqrt{2}} \tag{C.5}$$

and

$$\Psi_m(q) = \Psi_{asym}(q) \equiv \frac{\Psi(q) - \Psi(Cq)}{\sqrt{2}}, \tag{C.6}$$

where $\Psi(q)$ is a function that is localized near one of the potential minima and satisfies the Schrödinger equation up to exponentially small terms. Inserting this into Equation (C.4), we get

$$\Delta E = \frac{2\int dq \sqrt{g}\delta\left(f(q)\right)\frac{\partial f}{\partial q^i}\Psi\sqrt{g}\frac{\partial \Psi}{\partial q^j}}{\int dq\sqrt{g}\Psi^2}, \tag{C.7}$$

which is nothing but the Herring formula in general case.

D Derivation of Equation (6.97)

We consider the family of characteristics $[\mathbf{q}(\gamma), \mathbf{p}(\gamma)]$ for the Hamilton-Jacobi equation where γ is N-dimensional parameter. The matrix of the second derivatives $\tilde{A}_{ij}$ in Equation (6.96) is readily expressed in terms of the derivatives with respect to this parameter,

$$\tilde{\mathbf{A}} = \frac{D\mathbf{p}}{D\mathbf{q}} = \frac{D\mathbf{p}}{D\gamma}\left(\frac{D\mathbf{q}}{D\gamma}\right)^{-1}, \tag{D.1}$$

where we have used the notations $(D\mathbf{q}/D\gamma)_{ij} = \partial q_i/\partial\gamma_j$ and $(D\mathbf{p}/D\gamma)_{ij} = \partial p_i/\partial\gamma_j$. Along the characteristics, it satisfies the equation

$$\frac{d}{d\tau}\tilde{\mathbf{A}} = \left[\frac{d}{d\tau}\left(\frac{D\mathbf{p}}{D\gamma}\right)\right]\left(\frac{D\mathbf{q}}{D\gamma}\right)^{-1} - \tilde{\mathbf{A}}\left[\frac{d}{d\tau}\left(\frac{D\mathbf{q}}{D\gamma}\right)\right]\left(\frac{D\mathbf{q}}{D\gamma}\right)^{-1}. \tag{D.2}$$

Differentiating the Hamilton equations of motion for the characteristics with respect to γ, we get

$$\frac{d}{d\tau}\left(\frac{D\mathbf{p}}{D\gamma}\right) = -H_{\mathbf{qq}}\left(\frac{D\mathbf{q}}{D\gamma}\right) - H_{\mathbf{qp}}\left(\frac{D\mathbf{p}}{D\gamma}\right) \tag{D.3}$$

and

$$\frac{d}{d\tau}\left(\frac{D\mathbf{q}}{D\gamma}\right) = H_{\mathbf{pq}}\left(\frac{D\mathbf{q}}{D\gamma}\right) + H_{\mathbf{pp}}\left(\frac{D\mathbf{p}}{D\gamma}\right). \tag{D.4}$$

Inserting these relations into Equation (D.2), we get Equation (6.97) of the text.

E Computer Code to Calculate Instanton Trajectory

This appendix presents a Fortran source code for the variational calculations of the instanton path in the simplest case of NCOR Cartesian coordinates with the corresponding masses MASS(1:NCOR). One possible application is the tunneling problem in a large atomic cluster with frozen motion of distantly located atoms. At each step of iteration, the minimization of the classical action is done by using the simplest Newton's method, which involves the Hessian matrix of the potential function. Computing the Hessian can be avoided by switching to the conjugate gradient method. Note, however, that the Hessian matrix is ultimately needed in the final calculations of tunneling splitting and decay rate. In order to test the code, the following functions and subroutines should be provided:

1. Potential function POT(COR) of NCOR Cartesian coordinates COR(1: NCOR).
2. Subroutine GAULEG(XX,WW,NDVR): gives NDVR Gauss-Legendre quadrature points XX(1:NDVR) and the corresponding weights WW(1: NDVR).
3. Subroutine LEGPOL(Z,NFUN,FUN0,FUN1): gives NFUN lowest Legendre polynomials FUN0(1:NFUN) and their derivatives FUN1(1:NFUN). Z is the argument of the polynomials.
4. Subroutine PDER(COR,POT1,POT2,NCOR): gives NCOR first derivatives POT1(*) and NCOR*(NCOR+1)/2 second derivatives POT2(*) in symmetric storage mode.
5. Subroutine DIAG(A,N,E,Z,...): diagonalizer of real symmetric matrix A of order N in symmetric storage mode, whose eigenvalues and eigenvectors are E(N), Z(N,N).

```
      PROGRAM TRAJ
      PARAMETER (NCOR=30,NFUND=31, NGDIM=NFUND*NCOR,
     & NGDIMS=NGDIM*(NGDIM+1)/2, NCORS=NCOR*(NCOR+1)/2)
C
      DIMENSION POL(NFUND,NDIM),PR(NFUND,NDIM)
      DIMENSION XX(NDIM),WW(NDIM)
      DIMENSION S2(NGDIMS),S1(NGDIM),C(NGDIM)
      DIMENSION VAL(NGDIM),VEC(NGDIM,NGDIM)
      DIMENSION TR(NCOR,NDIM),DTR(NCOR,NDIM)
      DIMENSION TR0(NCOR,NDIM),DTR0(NCOR,NDIM)
      DIMENSION V(NDIM),VD(NCOR),VDD(NCORS)
      DIMENSION DNOR(NFUND),WK(NCOR,NCOR)
```

```
        DIMENSION XI(NCOR),XF(NCOR)
        REAL MASS(NCOR)
C — Input parameters
C     XI,XF - Cartesian coordinates of two potential minima (pot(xi)=pot(xf))
C     MASS - Atomic masses
C     NFUN - Number of basis functions (normally ~ 10-20);
C                 NFUN must not exceed NFUND
C     N     - Extra parameter: NDVR = N+NFUN becomes a number of quadrature
C                points in Gauss-Legendre integration, must not exceed NDIM
C     EPS    - Tolerance to cut small eigenvalues, also
C                used in the convergence criterion. The value depends on the
C                system. EPS = 1.E-3 could be a good try
C     Two other convergence criteria: (*) the 1st eigenvalue of δ^2 S_0 tends to zero;
C                                             (**) decreasing max(s1(1:ncor))
        namelist /input/xi,xf,mass,nfun,n,eps
        open(1,file='input.dat')          !Input data file
          read(1,input)
        close(1)
         ndvr=nfun+n
         ng=nfun*ncor                    !Total number of expansion coefficients in
Equation (6.111)
            if(ndvr.ge.ndim) stop 'INCREASE NDIM'
            if(nfun.ge.nfund) stop 'INCREASE NFUND'
         sqrt2=sqrt(2.0)

c — Legendre polynomials P_n(z) and their derivatives at quadrature points
        call gauleg(xx,ww,ndvr)
        do i=1,ndvr
         z=xx(i)
          call legpol(z,nfun,pol(1,i),pr(1,i))
        end do

c — Basis functions (1-z^2)P_n(z) and their derivatives at quadrature points
        do k=1,ndvr
         z=xx(k)
          z1=2*z
          z2=z*z
           do l=1,nfun
            pr(l,k)=pr(l,k)*(1.0-z2)-z1*pol(l,k)
            pol(l,k)=pol(l,k)*(1.0-z2)
           end do
        end do
        ezero = pot(xi)                  !Energy at the potential minimum
```

```
C — The initial straight path (first two terms in the right-hand side of Equation
(6.111))
C       TR0, DTR0: coordinates and their z derivatives at quadratures points
          do k=1,ndvr
            do ic=1,ncor
              tr0(ic,k)=0.5*(xi(ic)+xf(ic))+0.5*(xf(ic)-xi(ic))*xx(k)
              dtr0(ic,k)=0.5*(xf(ic)-xi(ic))
            end do
          end do

C — Initialization
          c=0.0
          niter=0
10        niter=niter+1                                   !Counting iterations
          if(niter.gt.50) stop 'NOT CONVERGED'            !It's time to look for a bug

C — Path and its z derivatives
          tr=tr0
          dtr=dtr0
            do k=1,ndvr
              ic=0
                do l=1,nfun
                  do i=1,ncor
                    ic=ic+1
                      tr(i,k)=tr(i,k)+C(ic)*pol(l,k)
                      dtr(i,k)=dtr(i,k)+C(ic)*pr(l,k)
                  end do
                end do
            end do

          s0=0.0
          s1=0.0
          s2=0.0
          dnor=0.0
C
          do 10000 k=1,ndvr                     !Loop over quadrature points
            sum=0.0
              do i=1,ncor
                sum=sum+dtr(i,k)*dtr(i,k)*mass(i)
              end do
            temp=sqrt(sum)
C — potential and its derivatives
            v(k)=pot(tr(1,k))-ezero
            call pder(tr(1,k),vd,vdd,ncor)
C — auxiliary normalization factors
```

```
          do i=1,nfun
            dnor(i)=dnor(i)+pol(i,k)*pol(i,k) * temp/sqrt(v(k))*ww(k)
          end do
C — Contribution to the action integral S0
          s0=s0+sqrt(v(k)) * temp * ww(k)
C — Contribution to δS0
          temp = sqrt2*sqrt(v(k))/temp
            do i=1,ncor
              wk(i,1)=dtr(i,k)*mass(i)
            end do

            ic=0
            do l=1,nfun
              do i=1,ncor
              ic=ic+1
                s1(ic)=s1(ic)+ww(k)*
     &         (temp*pr(l,k)*wk(i,1) + vd(i)*pol(l,k)/temp)
              end do
            end do
C — Contribution to δ²S0 (symmetric storage mode)
          do 100 ifun=1,nfun
            ig0=(ifun-1)*ncor
            do 110 jfun=1,ifun-1
              jg0=(jfun-1)*ncor
              do ic=ncor,1,-1
                ig=ig0+ic
                ij0=(ig*(ig-1))/2
                  do jc=1,ic
                     jg=jg0+jc
                       ij=ij0+jg
                         ijc=ic*(ic-1)/2+jc
                           wk(ic,jc)=vdd(ijc)*pol(ifun,k)*pol(jfun,k)/temp
                           s2(ij)=s2(ij)+ww(k)*wk(ic,jc)
                  enddo
                s2(ij)=s2(ij)+
     &                     ww(k)*temp*pr(ifun,k)*pr(jfun,k)*mass(ic)
                do jc=ic+1,ncor
                   jg=jg0+jc
                     ij=ij0+jg
                       s2(ij)=s2(ij)+ww(k)*wk(jc,ic)
                enddo
              enddo
110       continue
C — Diagonal block of δ²S0
          ijc=0
          do ic=1,ncor
```

```
              ig=ig0+ic
              ij0=(ig*(ig-1))/2
                do jc=1,ic
                  ijc=ijc+1
                    jg=ig0+jc
                      ij=ij0+jg
                    sum= vdd(ijc)*pol(ifun,k)*pol(jfun,k)/temp
                  s2(ij)=s2(ij)+ww(k)*sum
                enddo
C — Kinetic term in δ²S0
              sum=temp*pr(ifun,k)*pr(jfun,k)*mass(ic)
              s2(ij)=s2(ij)+ww(k)*sum
            enddo
100        continue
10000     continue
          s0=s0*sqrt2
C — Accuracy estimate
          del=abs(act-s0)
          if(del/s0.lt.eps) goto 1000          !The action is converged

C — Rescaling factors dnor(*) are introduced in order to have normalized
C      basis functions in Equation (6.113)
          do i=1,nfun
            dnor(i)=sqrt(sqrt2/dnor(i))
          end do
C — Rescaling δS0
          ic=0
          do l=1,nfun
            do i=1,ncor
              ic=ic+1
              s1(ic)=s1(ic)*dnor(l)
            end do
          end do
C — Rescaling δ²S0
          do ifun=1,nfun
            ig0=(ifun-1)*ncor
              do jfun=1,ifun-1
                jg0=(jfun-1)*ncor
                do ic=1,ncor
                  ig=ig0+ic
                  ij0=(ig*(ig-1))/2
                    do jc=1,ncor
                      jg=jg0+jc
                      ij=ij0+jg
                        s2(ij)=s2(ij)*dnor(ifun)*dnor(jfun)
                    enddo
```

```
          enddo
        enddo
        do ic=1,ncor
          ig=ig0+ic
          ij0=(ig*(ig-1))/2
            do jc=1,ic
              jg=ig0+jc
              ij=ij0+jg
                s2(ij)=s2(ij)*dnor(ifun)*dnor(ifun)
            enddo
        enddo
      enddo
C — Diagonalization of δ²S0
      call diag(s2,ng,val,vec,...)
C     val(1:ng), vec(1:ng,1:ng) are assumed to contain eigenvalues and eigenstates
C     The negative eigenvalues are omitted.
C     For convergent instanton path val(1) → 0 and val(2:ncor) > 0.

      s2(1:ng)=matmul(transpose(vec(1:ng,1:ng)),s1(1:ng))
      do k=2,ng
        if(val(k).gt.eps)then
          s1(k)=-s2(k)/val(k)
        else
          s1(k)=0.0
        end if
      end do
C — Calculate the new expansion coefficients
      ic=0
        do l=1,nfun
          do i=1,ncor
            ic=ic+1
            sum=0.0
              do k=2,ng
                sum=sum+vec(ic,k)*s1(k)
              end do
            c(ic)=c(ic)+sum*dnor(l)
          end do
        end do
C — Estimate new S0
      act=s0
      do k=2,ng
        if(val(k).gt.eps) act=act-0.5*s2(k)**2/val(k)
      end do
      goto 10                !Proceed with the next iteration
1000  continue               !Done; C(1:ng) determine the instanton path Equa-
tion (6.111)
      end
```

As an illustration, we present the results of the iterative calculations of S_0 in the model potential

$$V(\{x\}) = (x_1^2 - 1)^2 + \sum_{i=2}^{N_c} \left(\frac{x_i^2}{2} + x_1 x_i \right) \tag{E.1}$$

of $N_c = 10$ Cartesian coordinates with unit mass factors, computed by using 10 and 20 basis functions in Equation (6.111) (parameter NFUN in the code). Two other input parameters were taken as N=10 and EPS=1.E-4. In practice, the first few iterations always give physical result with two to three significant digits of accuracy. Two above-mentioned extra convergence criteria are also shown in the tables. These data can be directly used in the code compilation.

NFUN=10

N_{iter}	S_0	S1MAX	VAL(1)
1	34.93644	12.13063	-2.335472
2	29.72368	5.465971	0.146039
3	28.95526	0.361709	2.071044E-002
4	28.95110	1.169700E-002	7.401369E-003
5	28.95109	8.930430E-003	7.342669E-003

NFUN=20

N_{iter}	S_0	S1MAX	VAL(1)
1	34.93644	12.13063	9.179749E-002
2	30.45321	4.866121	8.4116373E-002
3	29.41160	1.748358	8.177740E-002
4	29.12340	0.9158290	7.722782E-002
5	29.03844	0.7215158	8.080321E-002
6	29.00676	0.5661420	8.140273E-002
7	28.99099	0.4374829	8.103709E-002
8	28.98333	0.3416091	8.082822E-002
9	28.97825	0.3145442	8.084504E-002
10	28.97465	0.2593439	8.048192E-002
11	28.97201	0.3137331	8.025333E-002
12	28.96978	0.2365213	8.010309E-002
13	28.96796	0.2631376	7.999041E-002
14	28.96641	0.2115241	7.974611E-002

F Derivation of Some Equations in Section 6.4.2

F.1 Derivation of Equation (6.156)

Since Equation (6.153) explicitly defines the function $\tau(\mathbf{x})$ and $\mathbf{v}_0 = \partial \mathbf{x}_0 \partial \tau$, we have

$$\frac{\partial(\Delta \mathbf{x} \mathbf{v}_0)}{\partial \mathbf{x}} \equiv \frac{\partial[(\mathbf{x}-\mathbf{x}_0)\mathbf{v}_0]}{\partial \mathbf{x}} = \mathbf{v}_0 + \left(\Delta \mathbf{x} \dot{\mathbf{v}}_0 - \mathbf{v}_0^2\right) \frac{\partial \tau}{\partial \mathbf{x}} = 0. \tag{F.1}$$

Equation (6.156) follows directly from this equation. The second equal in Equation (6.156) is just an expansion with respect to $\Delta \mathbf{x}$.

F.2 Derivation of Equation (6.157)

Equation (6.155) gives

$$\frac{d(\mathbf{a} - \mathbf{A}\mathbf{v}_0)}{d\tau} = V_{\mathbf{xx}} \mathbf{v}_0 - \dot{\mathbf{A}} \mathbf{v}_0 - \mathbf{A}\mathbf{a} = -\mathbf{A}(\mathbf{a} - \mathbf{A}\mathbf{v}_0), \tag{F.2}$$

which is a system of linear differential equations with zero initial condition at $\tau \to -\infty$ (potential minimum), $\mathbf{a}(-\infty) = \nabla V(-\infty) = 0$. Thus $\mathbf{a} - \mathbf{A}\mathbf{v}_0 = 0$ holds at all τ, which is Equation (6.157).

F.3 Derivation of Equation (6.154)

Since we have

$$\frac{\partial}{\partial \mathbf{x}} \sum_i \int_{-\infty}^{\tau} p_{0i}(\tau') \dot{x}_{0i}(\tau') d\tau' = \frac{\partial \tau}{\partial \mathbf{x}} \sum_i p_{0i}(\tau) \dot{x}_{0i}(\tau) = \frac{\partial \tau}{\partial \mathbf{x}} \mathbf{v}_0^2 \tag{F.3}$$

and

$$\frac{\partial}{\partial \mathbf{x}} \sum_{ij} A_{ij} [x_i - x_{0i}][x_j - x_{0j}] = \frac{\partial}{\partial \mathbf{x}} (\Delta \mathbf{x}^T \mathbf{A} \Delta \mathbf{x})$$
$$= \frac{\partial \tau}{\partial \mathbf{x}} \Delta \mathbf{x}^T \dot{\mathbf{A}} \Delta \mathbf{x} + \mathbf{A} \Delta \mathbf{x} - \Delta \mathbf{x}^T \mathbf{A} \frac{\partial \tau}{\partial \mathbf{x}} \mathbf{v}_0, \tag{F.4}$$

the differentiation of Equation (6.154) with respect to $\mathbf{x}$ leads to

$$\begin{aligned}\frac{\partial W_0}{\partial \mathbf{x}} &= \mathbf{v}_0\left(1+\frac{(\mathbf{a}\Delta\mathbf{x})}{\mathbf{v}_0{}^2}+\frac{(\mathbf{a}\Delta\mathbf{x})^2}{(\mathbf{v}_0^2)^2}\right)+\frac{1}{2}\left(\Delta\mathbf{x}^T V_{\mathbf{xx}}\Delta\mathbf{x}\right)\frac{\mathbf{v}_0}{\mathbf{v}_0^2}-\frac{1}{2}\left(\Delta\mathbf{x}^T\mathbf{A}^2\Delta\mathbf{x}\right)\frac{\mathbf{v}_0}{\mathbf{v}_0^2}\\ &\quad+\mathbf{A}\Delta\mathbf{x}-\frac{\mathbf{v}_0}{\mathbf{v}_0^2}\left(\Delta\mathbf{x}^T\mathbf{A}\mathbf{v}_0\right)\left(1+\frac{\mathbf{a}\Delta\mathbf{x}}{\mathbf{v}_0^2}\right)+o((\Delta\mathbf{x})^2)\\ &= \mathbf{v}_0+\mathbf{A}\Delta\mathbf{x}+\frac{1}{2}\left(\Delta\mathbf{x}^T V_{\mathbf{xx}}\Delta\mathbf{x}\right)\frac{\mathbf{v}_0}{\mathbf{v}_0^2}-\frac{1}{2}\left(\Delta\mathbf{x}^T\mathbf{A}^2\Delta\mathbf{x}\right)\frac{\mathbf{v}_0}{\mathbf{v}_0^2}+o((\Delta\mathbf{x})^2), \quad \text{(F.5)}\end{aligned}$$

where Equations (6.156) and (6.157) have been used. Then we have

$$\begin{aligned}\frac{1}{2}\left(\frac{\partial W_0}{\partial \mathbf{x}}\right)^2 &= \frac{\mathbf{v}_0^2}{2}+\mathbf{v}_0^T\mathbf{A}\Delta\mathbf{x}+\frac{1}{2}\Delta\mathbf{x}^T V_{\mathbf{xx}}\Delta\mathbf{x}-\frac{1}{2}\Delta\mathbf{x}^T\mathbf{A}^2\Delta\mathbf{x}\\ &\quad+\frac{1}{2}(\mathbf{A}\Delta\mathbf{x})^T(\mathbf{A}\Delta\mathbf{x})+o((\Delta\mathbf{x})^2)\\ &= V(\mathbf{x}_0)+V_{\mathbf{x}}\Delta\mathbf{x}+\frac{1}{2}\Delta\mathbf{x}^T V_{\mathbf{xx}}\Delta\mathbf{x}+o((\Delta\mathbf{x})^2), \quad \text{(F.6)}\end{aligned}$$

where $\mathbf{v}_0 = \mathbf{p}_0 = \sqrt{2V}$ and $\mathbf{a} = \mathbf{A}\mathbf{v}_0 = \nabla V$ are used. This means that the Hamilton-Jacobi equation in the vicinity of the instanton path is satisfied up to the orders $o[(\Delta\mathbf{x})^2]$.

F.4 Derivation of Equation (6.162)

From Equation (6.161) we have

$$\begin{aligned}\frac{\partial w}{\partial \mathbf{x}} &= \frac{\partial \tau}{\partial \mathbf{x}}[\theta(\tau)-\Delta E]-\frac{1}{\mathbf{U}^T\Delta\mathbf{x}}\frac{\partial}{\partial \mathbf{x}}[\mathbf{U}^T\Delta\mathbf{x}]\\ &= \frac{\mathbf{v}_0}{\mathbf{v}_0^2}\left(1+\frac{\mathbf{a}\Delta\mathbf{x}}{\mathbf{v}_0^2}+\frac{(\mathbf{a}\Delta\mathbf{x})^2}{(\mathbf{v}_0^2)^2}\right)(\theta-\Delta E)\\ &\quad-\frac{1}{\mathbf{U}^T\Delta\mathbf{x}}\left[\frac{\partial \tau}{\partial \mathbf{x}}\frac{d\mathbf{U}^T}{d\tau}\Delta\mathbf{x}+\mathbf{U}^T-\frac{\partial \tau}{\partial \mathbf{x}}\mathbf{U}^T\mathbf{v}_0\right]\\ &= \frac{\mathbf{v}_0}{\mathbf{v}_0^2}\left(1+\frac{\mathbf{a}\Delta\mathbf{x}}{\mathbf{v}_0^2}\right)(\theta-\Delta E)-\frac{1}{\mathbf{U}^T\Delta\mathbf{x}}\left[\dot{\mathbf{U}}^T\Delta\mathbf{x}\frac{\mathbf{v}_0}{\mathbf{v}_0^2}+\frac{\mathbf{v}_0}{\mathbf{v}_0^2}\frac{\mathbf{a}\Delta\mathbf{x}}{\mathbf{v}_0^2}\dot{\mathbf{U}}^T\Delta\mathbf{x}+\mathbf{U}^T\right.\\ &\quad\left.-\frac{\mathbf{v}_0}{\mathbf{v}_0^2}\left(1+\frac{\mathbf{a}\Delta\mathbf{x}}{\mathbf{v}_0^2}+\frac{(\mathbf{a}\Delta\mathbf{x})^2}{(\mathbf{v}_0^2)^2}\right)(\mathbf{U}^T\mathbf{v}_0)\right]+O\left((\Delta\mathbf{x})^2\right), \quad \text{(F.7)}\end{aligned}$$

where Equation (6.156) is used. Multiplying $\partial W_0/\partial \mathbf{x}\left[= \mathbf{v}_0 + \mathbf{A}\Delta\mathbf{x} + O\left((\Delta\mathbf{x})^2\right)\right]$ from Equation (F.5), we have

$$
\begin{aligned}
(\mathbf{U}^T\Delta\mathbf{x})\frac{\partial W_0}{\partial \mathbf{x}}\frac{\partial w}{\partial \mathbf{x}} &= (\mathbf{U}^T\Delta\mathbf{x})(\mathbf{v}_0 + \mathbf{A}\Delta\mathbf{x})\frac{\mathbf{v}_0}{\mathbf{v}_0^2}\left(1 + \frac{\mathbf{a}\Delta\mathbf{x}}{\mathbf{v}_0^2}\right)(\theta - \Delta E) - (\mathbf{v}_0 + \mathbf{A}\Delta\mathbf{x}) \\
&\quad \times \left[\dot{\mathbf{U}}^T\Delta\mathbf{x}\frac{\mathbf{v}_0}{\mathbf{v}_0^2} + \frac{\mathbf{v}_0}{\mathbf{v}_0^2}\frac{\mathbf{a}\Delta\mathbf{x}}{\mathbf{v}_0^2}\dot{\mathbf{U}}^T\Delta\mathbf{x} + \mathbf{U}^T - \frac{\mathbf{v}_0}{\mathbf{v}_0^2}\left(1 + \frac{\mathbf{a}\Delta\mathbf{x}}{\mathbf{v}_0^2}\right)(\mathbf{U}^T\mathbf{v}_0)\right] \\
&= (\mathbf{U}^T\Delta\mathbf{x})(\theta - \Delta E) - \left[\dot{\mathbf{U}}^T\Delta\mathbf{x} + \mathbf{U}^T\mathbf{A}\Delta\mathbf{x}\right. \\
&\quad \left. - \frac{\mathbf{v}_0}{\mathbf{v}_0^2}\mathbf{A}\Delta\mathbf{x}\mathbf{U}^T\mathbf{v}_0 - \frac{\mathbf{a}\Delta\mathbf{x}}{\mathbf{v}_0^2}\mathbf{U}^T\mathbf{v}_0\right] + O\left((\Delta\mathbf{x})^2\right). \qquad \text{(F.8)}
\end{aligned}
$$

Thus from Equation (6.159) we get

$$
\begin{aligned}
\left[\frac{\partial W_0}{\partial \mathbf{x}}\frac{\partial w}{\partial \mathbf{x}} + \Delta E\right](\mathbf{U}^T\Delta\mathbf{x}) &= \mathbf{U}^T\Delta\mathbf{x}\theta - \dot{\mathbf{U}}^T\Delta\mathbf{x} - \mathbf{U}^T\mathbf{A}\Delta\mathbf{x} \\
&\quad + 2\frac{\mathbf{A}\mathbf{v}_0\Delta\mathbf{x}}{\mathbf{v}_0^2}\mathbf{U}^T\mathbf{v}_0 + O\left((\Delta\mathbf{x})^2\right) = 0, \qquad \text{(F.9)}
\end{aligned}
$$

where Equation (6.157) is used. This gives Equation (6.162).

F.5 Consistency of Equations (6.163) and (6.165) with Equation (6.167)

In the case of the orthogonality condition, Equation (6.163), with use of Equations (6.167) we obtain

$$
\begin{aligned}
\frac{d}{d\tau}(\mathbf{U}^T\mathbf{v}_0) &= \dot{\mathbf{U}}^T\mathbf{v}_0 + \mathbf{U}^T\dot{\mathbf{v}}_0 = (\mathbf{U}^T\mathbf{A}\mathbf{U})(\mathbf{U}^T\mathbf{v}_0) - \mathbf{U}^T\mathbf{A}\mathbf{v}_0 + \mathbf{U}^T\mathbf{a} \\
&= (\mathbf{U}^T\mathbf{A}\mathbf{U})(\mathbf{U}^T\mathbf{v}_0) + \mathbf{U}^T(\mathbf{a} - \mathbf{A}\mathbf{v}_0) = 0. \qquad \text{(F.10)}
\end{aligned}
$$

The normalization condition, Equation (6.165), can be easily confirmed with use of Equation (6.167).

Bibliography

1. T. Banks, C.M. Bender, and T.T. Wu, *Phys. Rev.* **D8**, 3346 (1973); T. Banks and C.M. Bender, *Phys. Rev.* **D8**, 3366 (1973).
2. S. Coleman, "The Use of Instanton," in *The Whys of Subnuclear Physics* edited by A. Zichichi (Plenum, New York, 1979).
3. C.G. Callan and S. Coleman, *Phys. Rev.* **D16**, 1762 (1977).
4. A. Auerbach and S. Kivelson, *Nucl. Phys.* **B257**, 799 (1985).
5. A. Iwamoto and Y. Tomita, *Prog. Theo. Phys.* **87**, 1171 1992).
6. A.O. Caldeira and A.J. Leggett, *Ann. Phys. (NY)* **149**, 374 (1983); **153**, 445 (E) (1984).
7. Z.H. Huang, T.E. Feuchtwang, P.H. Cutler, and E. Kazes, *Phys. Rev.* **A41**, 32 (1990).
8. H. Haug and A.-P. Jauho, *Quantum Kinetics in Transport and Optics of Semiconductors* (Springer, Berlin, 1998).
9. R.P. Bell, *Tunneling Effects in Chemistry* (Chapman and Hall, London, 1980).
10. D.G. Truhlar and B.C. Garrett, *Annu. Rev. Phys. Chem.* **35**, 159 (1984).
11. D.G. Truhlar, A.D. Isaacson, and B.C. Garrett, *Theory of Chemical Reaction Dynamics* edited by M. Baer (CRC, Boca Raton, 1985), Chap. 2.
12. J. Jortner and B. Pullman, *Tunneling* (Reidel, Dordrecht, 1986).
13. V.I. Goldanskii, L.I. Trakhtenberg, and V.N. Fleurov, *Tunneling Phenomena in Chemical Physics* (Gordon and Breach, New York, 1989).
14. H. Nakamura, *Int. Rev. Phys. Chem.* **10**, 123 (1991).
15. S. Takada, *J. Chem. Phys.* **104**, 3742 (1996).
16. V.I. Goldanskii and J. Jortner (eds.), *Special issue on Tunneling in Chemical Reactions, Chem. Phys.* **170** (1993).
17. V.A. Benderskii, D.E. Makarov, and C.A. Wight, *Chemical Dynamics at Low Temperatures* (Advances in Chemical Physics, John Wiley and Sons, New York, 1994).
18. V.A. Benderskii, V.I. Goldanskii, and D.E. Makarov, *Phys. Rep.* **233**, 195 (1993).
19. B. Chance, D.C. Devault, H. Frauenfelder, R.A. Marcus, J.R. Schriefer, and N. Sutin, *Tunneling in Biological Systems* (Academic, New York, 1979).
20. D. Devault, *Quantum Mechanical Tunneling in Biological Systems* (Cambridge University Press, Cambridge, 1984).
21. M. Razavy, *Quantum Theory of Tunneling* (World Scientific, Singapore, 2003).
22. T. Miyazaki (ed.), *Atom Tunneling Phenomena in Physics, Chemistry and Biology* (Springer, Berlin, 2004).
23. J. Ankerhold, *Quantum Tunneling in Complex Systems: The Semiclassical Approach* (Springer, Berlin, 2007).
24. S. Keshavamurthy and P. Schlagheck (eds.), *Dynamical Tunneling: Theory and Experiment* (Taylor and Francis, Boca Raton, 2011)
25. H.-D. Meyer, F. Gatti, and G.A. Worth (eds.), *Multidimensional Quantum Dynamics: MCTDH Theory and Applications* (Wiley-VCH, Weinheim, 2009).
26. J.B. Anderson (ed.), *Quantum Monte Carlo: Origins, Development, Applications* (Oxford University Press, New York, 2007).
27. W.H. Miller, *J. Chem. Phys.* **62**, 1899 (1975).
28. F. Cesci, G.C. Rossi, and M. Testa, *Ann. Phys.* **206**, 318 (1991).
29. M. Wilkinson, *Physica* **D21**, 341 (1986).
30. S. Takada and H. Nakamura, *J. Chem. Phys.* **100**, 98 (1994); *Supp. Prog. Theo. Phys.* **116**, 295 (1994).
31. S. Takada and H. Nakamura, *J. Chem. Phys.* **102**, 3977 (1995).

32. P. Bowcock and R. Gregory, *Phys. Rev.* **D44**, 1774 (1991).
33. A. Schmid, *Ann. Phys.(NY)* **170**, 333 (1986).
34. (a) V.A. Benderskii, E.V. Vetoshkin, S.Yu. Grebenschikov, L. von Laue, and H.P. Trommsdorff, *Chem. Phys.* **219**, 119 (1997); (b) V.A. Benderskii, S.Yu. Grebenschikov, and G.V. Mil'nikov, *Chem. Phys.* **198**, 281 (1995); (c) Ibid. **194**, 1 (1995); (d) V.A. Benderskii, S.Yu. Grebenschikov, D.E. Makarov, and E.V. Vetoshkin, *Chem. Phys.* **185**, 101 (1994).
35. N. Makri and W.H. Miller, *J. Chem. Phys.* **91**, 4026 (1989); *J. Chem. Phys.* **86**, 1451, **87**, 5781 (1987); W.H. Miller, *J. Chem. Phys.* **55**, 3146 (1971); **62**, 1899 (1975); *J. Chem. Phys.* **105**, 2942 (2001).
36. K. Takatsuka, H. Ushiyama, and A. Inoue-Ushiyama, *Phys. Rep.* **322**, 347 (1999).
37. G.V. Mil'nikov and A.J. Varandas, *J. Chem. Phys.* **111**, 8302 (1999).
38. S. Coleman, *Phys. Rev.* **D15**, 2929 (1977).
39. A.I. Vainshtein, V.I. Zakharov, V.A. Novikov, and M.A. Shifman, *Sov. Phys. Usp* **25**, 195 (1982).
40. V.A. Benderskii, E.V. Vetoshkin, L. von Laue, and H.P. Trommsdorff, *Chem. Phys.* **219**, 143 (1997); V.A. Benderskii, E.V. Vetoshkin, and H.P. Trommsdorff, *Chem. Phys.* **244**, 273 (1999); V.A. Benderskii, I.S. Irgibaeva, E.V. Vetoshkin, and H.P. Trommsdorff, *Chem. Phys.* **262**, 369 (2000); V.A. Benderskii, E.V. Vetoshkin, I.S. Irgibaeva, and H.P. Trommsdorff, *Chem. Phys.* **262**, 393 (2000).
41. Z. Smedarchina, W. Siebrand, and M.Z. Zgierski, *J. Chem. Phys.* **103**, 5326 (1995); J.P. Roscioli, D.W. Pratt, Z. Smedarchina, W. Siebrand, and A. Fernandez-Ramo, *J. Chem. Phys.* **120**, 11351 (2004); Z. Smedarchina, W. Siebrand, M.Z. Zgierski, and F. Zerbetto, *J. Chem. Phys.* **103**, 5326 (1995); F. Fernandez-Ramos, Z. Smerdarchina, and J. Rodriguez-Otero, *J. Chem. Phys.* **114**, 1567 (2001); Z. Smedarchina and W. Siebrand, *Chem. Phys.* **170**, 347 (1993); Z. Smedarchina, A. Fernandez-Ramos, and W. Siebrand, *J. Chem. Phys.* **122**, 134309 (2005); Z. Smedarchina and W. Siebrand, *Chem. Phys. Lett.* **395**, 339 (2004); Z. Smedarchina, W. Siebrand, A. Fernandez-Ramos, and E. Martinez-Nunez, *Chem. Phys. Lett.* **386**, 396 (2004); Z. Smedarchina, W. Siebrand, and M.Z. Zgierski, *J. Chem. Phys.* **104**, 1203 (1996).
42. C.S. Tautermann, A.F. Voegele, T. Loerting, and K.R. Liedl, *J. Chem. Phys.* **117**, 1962 (2002); C.S. Tauterman, A.F. Voegele, T. Loerting, and K.R. Liedl, *J. Chem. Phys.* **117**, 1967 (2002); C.S. Tauterman, A.F. Voegele, and K.R. Liedl, *J. Chem. Phys.* **120**, 631 (2004).
43. G.V. Mil'nikov and H. Nakamura, *J. Chem. Phys.* **115**, 6881 (2001).
44. G.V. Mil'nikov and H. Nakamura, *J. Chem. Phys.* **117**, 10081 (2002).
45. G.V. Mil'nikov and H. Nakamura, *J. Chem. Phys.* **122**, 124311 (2005).
46. G.V. Mil'nikov and H. Nakamura, *Phy. Chem. Chem. Phys.* **10**, 1374 (2008).
47. P. Oloyede, G.V. Mil'nikov, and H. Nakamura, *J. Theo. Comp. Chem.* **3**, 91 (2004).
48. H. Nakamura, *Nonadiabatic Transition: Concepts, Basic Theories and Applications* [2nd edition] (World Scientific, Singapore, 2012).
49. C. Eckart, *Phys. Rev.* **35**, 1303 (1930).
50. L.D. Landau and E.M. Lifshitz, *Quantum Mechanics* (Pergamon Press, N.Y., 1981).
51. S.C. Miller, Jr. and R.H. Good, Jr., *Phys. Rev.* **91**, 174 (1953).
52. M.S. Child, *Semiclassical Mechanics with Molecular Applications* (Clarendon, Oxford, 1991).
53. G.V. Mil'nikov, S.Yu. Grebenshikov, and V.A. Benderskii, *Izv. Akad. Nauk (Russian Chem. Bull.)* No.12, 2098 (1994); V.A. Benderskii, S.Yu. Grebenshikov, and G.V. Mil'nikov, *Chem. Phys.* **194**, 1 (1995).
54. C. Herring, *Rev. Mod. Phys.* **34**, 631 (1962).
55. W.H. Miller, *J. Chem. Phys.* **83**, 960 (1979).

56. E. Bosch, M. Moreno, J.M. Lluch, and J. Bertan, *J. Chem. Phys.* **93**, 5685 (1990).
57. A.J. Lichtenberg and M.A. Lieberman, *Regular and Stochastic Motion* (Springer-Verlag, New York, 1983).
58. A.M. Ozorio de Almeida, *Hamiltonian Systems* (Cambridge University Press, Cambridge, 1988).
59. V.P. Maslov and M.V. Fedoriuk, *Semi-Classical Approximation in Quantum Mechanics* (Reidel, Dordrecht, Holland,1981).
60. J.B. Delos, *Adv. Chem. Phys.* **65**, 161 (1986).
61. S.K. Knudson, J.B. Delos, and D.W. Noid, *J. Chem. Phys.* **84**, 6886 (1986).
62. R. Courant and D. Hilbert, *Methods of Mathematical Physics. Vol.1* (Interscience, New York, 1953).
63. R.A. Marcus and M.E. Coltrin, *J. Chem. Phys.* **67**, 2609 (1977).
64. M.Ya. Ovchinikova, *Chem. Phys.* **36**, 85 (1979).
65. V.K. Babamov and R.A. Marcus, *J. Chem. Phys.* **74**, 1790).
66. H. Nakamura and A. Ohsaki, *J. Chem. Phys.* **83**, 1599 (1985); H. Nakamura, *Chem. Phys. Lett.* **141**, 77 (1987); A. Ohsaki and H. Nakamura, *Chem. Phys. Lett.* **142**, 37 (1987).
67. G.C. Garrett, D.G. Truhlar, A.F. Wagner, and T.H. Dunning, Jr., *J. Chem. Phys.* **78**, 4400 (1983).
68. J.P. Sethna, *Phys. Rev.* **B24**, 692 (1981).
69. N.Takigawa, K. Hagino, M. Abe, and A.B. Balentikin, *Phys. Rev.* **C49**, 2636 (1994).
70. R.L. Redington, Y. Chen, G.J. Scherer, and R.W. Field, *J. Chem. Phys.* **88**, 627 (1988).
71. Y. Ikegami, *Bull. Chem. Soc. Jpn.* **34**, 94 (1960); **36**, 1118 (196).
72. R.L. Redington and T.E. Redington, *J. Mol. Spectr.* **78**, 229 (1979).
73. K. Tanaka, H. Honjo, T. Tanaka, H. Kohguchi, Y. Ohshima, and Y. Endo, *J. Chem. Phys.* **110**, 1969 (1999).
74. Y. Tomioka, M. Ito, and N. Mikami, *J. Phys. Chem.* **87**, 4401 (1983).
75. A.C.P. Alves, J.M. Hollas, H. Musa, and T. Ridley, *J. Mol. Spectr.* **109**, 99 (1985).
76. H. Sekiya, Y. Nagashima, and Y. Nishimura, *J. Chem. Phys.* **92**, 5761 (1990); *Bull. Chem. Soc. Jpn.* **62**, 3229 (1989).
77. A.C.P. Alves and J.M. Hollas, *Mol. Phys.* **23**, 927 (1972); **25**, 1305 (1973).
78. R. Rossetti and L.E. Brus, *J. Chem. Phys.* **73**, 1546 (1980).
79. R.L. Redington, T.E. Redington, M.A. Hunter, and R.W. Field, *J. Chem. Phys.* **92**, 6456 (1990).
80. H. Sekiya, K. Sasaki, Y. Nishimura, Z. Li, A. Mori, and H. Takeshita, *Chem. Phys. Lett.* **173**, 285 (1990); H. Sekiya, Y. Nagashima, T. Tsuji, Y. Nishimura, A. Mori, and H. Takeshita, *J. Phys. Chem.* **95**, 10311 (1991).
81. A.E. Bracamonte and P.H. Vaccaro, *J. Chem. Phys.* **120**, 4638 (2004).
82. GAUSSIAN 92, Revision A.M. J. Frisch, G.W. Trucks, M. Head-Gordon, P.M.W. Gill, W. Wong, J.B. Foresman, B.G. Johnson, H.B. Schlegel, M.A. Robb, E.S. Replogle, R. Gomperts, J.L. Andres, K. Raghavachari, J.S. Binkley, C. Gonzalez, R.L. Martin, D.J. Fox, D.J. Defrees, J. Baker, J.J.P. Stewart, and J.A. Pople, Gaussian, Inc., Pittsburgh PA, 1992.
83. J.C. Light, I.P. Hamilton, and J.V. Lill, *J. Chem. Phys.* **82**, 1400 (1985); R.M. Whitnell and J.C. Light, ibid. **90**, 1774 (1989).
84. M.V. Verner, S. Scheiner, and N.D. Sokorov, *J. Chem. Phys.* **101**, 9755 (1994).
85. M.J. Wojcik, H. Nakamura, S. Iwata, and W. Tatara, *J. Chem. Phys.* **112**, 6322 (2000).
86. R. Casadesus, O. Vendrell, M. Moreno, and J.M. Lluch, *Chem. Phys. Lett.* **405**, 187 (2005).
87. M. Wojcik, L. Boda, and M. Boczar, *J. Chem. Phys.* **130**, 164306 (2009).

88. W. Domcke, D. Yarkony, and H. Koppel, *Conical Intersections, Theory, Computation and Experiment* (World Scientific, Singapore, 2011).
89. H.A. Jahn and E. Teller, *Proc. Roy. Soc.* **A161**, 220 (1937).
90. R. Englman, *The Jahn-Teller Effect in Molecules and Crystals* (Wiley-Interscience, New York, 1972).
91. I.B. Bersuker, *The Jahn-Teller Effect and Vibronic Interaction in Modern Chemistry* (Plenum, New York, 1984).
92. S. Nanbu, H. Nakamura, and F.O. Goodman, *J. Chem. Phys.* **107**, 5445 (1997).
93. H. Nakamura, *J. Chem. Phys.* **97**, 256 (1992).
94. H. Nakamura, *J. Chem. Phys.* **110**, 10253 (1999).
95. S. Nanbu, T. Ishida, and H. Nakamura, *Chem. Phys.* **324**, 721 (2006).
96. H. Zhang, S.C. Smith, S. Nanbu, and H. Nakamura, *J. Phys. Condens. Matter* **21**, 144209 (2009); *Austrlian J. Chem.* **63**, 371 (2010).
97. K. Nagaya, Y. Teranishi, and H. Nakamura, *J. Chem. Phys.* **113**, 6197 (2000).
98. J. Fujisaki, Y. Teranishi, and H. Nakamura, *J. Theo. Comp. Chem.* **1**, 245 (2002).
99. L.D. Landau, *Phys. Zts. Sov.* **2**, 46 (1932).
100. C. Zener, *Proc. Roy. Soc.* **A137**, 696 (1932).
101. E.C.G. Stückelberg, *Helv. Phys. Acta.* **5**, 369 (1932).
102. (1) C. Zhu and H. Nakamura, *J. Math. Phys.* **33**, 2697 (1992), (2) C.Zhu, H. Nakamura, N. Re, and V. Aquilanti, *J. Chem. Phys.* **97**, 1892 (1992), (3) C. Zhu and H. Nakamura, *J. Chem. Phys.* **97**, 8497 (1992), (4) ibid., **98**, 6208 (1993), (5) ibid., **101**, 4855 (1994), (6) ibid., **101**, 10630 (1994), (7) ibid., **102**, 7448 (1995), (8) ibid., **107**, 7839 (1997), (9) ibid., **108**, 7501 (1998), (10) ibid., **109**, 4689 (1998).
103. H. Nakamura, *Adv. Chem. Phys.* **82**, 243 (1992).
104. H. Nakamura, *Dynamics of Molecules and Chemical Reactions* edited by R.E. Wyatt and J.Z.H. Zhang (Marcel Dekker, New York, 1996), chap. 12.
105. H. Nakamura and C. Zhu, *Comm. Atom. and Molec. Phys.* **32**, 249 (1996).
106. H. Nakamura, *Ann. Rev. Phys. Chem.* **48**, 299 (1997).
107. C. Zhu, Y. Teranishi, and H. Nakamura, *Adv. Chem. Phys.* **117**, 127 (2001).
108. H. Nakamura, *Adv. Chem. Phys.* **138**, 95 (2008); H. Nakamura, *J. Phys. Chem.* **A110**, 10929 (2006).
109. (1) C. Zhu and H. Nakamura, *Chem. Phys. Lett.* **258**, 342 (1996), (2) ibid., *J. Chem. Phys.* **106**, 2599 (1997), (3) ibid., **107**, 7839 (1997), (4) ibid., *Chem. Phys. Lett.* **274**, 205 (1997).
110. P.S. Julienne and M.J. Krauss, *J. Mol. Spectr.* **56**, 270 (1975).
111. P. Oloyede, G.V. Mil'nikov, and H. Nakamura, *J. Chem. Phys.* **124**, 144110 (2006).
112. S. Nanbu, T. Ishida, and H. Nakamura, *Chemical Science* **1**, 663 (2010).
113. J.C. Tully and R.K. Preston, *J. Chem. Phys.* **54**, 4297 (1970); ibid., **55**, 562 (1971); J.C. Tully, *J. Chem. Phys.* **93**, 1061 (1990).
114. J.C. Tully, *Dynamics of Molecular Collisions, Part B* edited by W.H. Miller (Plenum, New York, 1976), p. 217.
115. J.C. Tully, *Modern Methods for Multidimensional Dynamics Computations in Chemistry* edited by D.L. Thompson (World Scientific, Singapore, 1998), p. 34.
116. M.D. Hack, A.W. Jasper, Y.L. Volobuev, D.W. Schwenke, and D.G. Truhlar, *J. Phys. Chem.* **103**, 6309 (1999).
117. A. Jasper, B.K. Kendrick, C.A. Mead, and D.G. Truhlar, in *Modern Trends in Chemical Reaction Dynamics: Experiment and Theory* edited by K. Liu and X. Yang (World Scientific, Singapore, 2004).
118. A. Kondorskii and H. Nakamura, *J. Chem. Phys.* **120**, 8937 (2004); ibid., *J. Theo. Comp. Chem.* **4**, 89 (2005).

119. M.F. Herman and E. Kluk, *Chem. Phys.* **91**, 27 (1984); E. Kluk, M.F. Herman, and H.L. Davis, *J. Chem. Phys.* **84**, 326 (1986); M.F. Herman, *Annu. Rev. Phys. Chem.* **45**, 83 (1994); M.F. Herman, *Chem. Phys. Lett.* **275**, 445 (1997).
120. A.R. Walton and D.E. Manolopoulos, *Mol. Phys.* **87**, 961 (1996).
121. W.H. Miller, *J. Chem. Phys.* **53**, 3578 (1970); X. Sun and W.H. Miller, *J. Chem. Phys.* **106**, 6346 (1997); W.H. Miller, *J. Phys. Chem.* **105**, 2942 (2001); ibid., *Proc. Nat. Acad. Sci.* **102**, 6660 (2005); ibid., *J. Chem. Phys.* **125**, 132305 (2006).
122. G.V. Mil'nikov, K. Yagi, T. Taketsugu, H. Nakamura, and K. Hirao, *J. Chem. Phys.* **120**, 5036 (2004).
123. G.V. Mil'nikov, K. Yagi, T. Taketsugu, H. Nakamura, and K. Hirao, *J. Chem. Phys.* **119**, 10 (2003).
124. T.H. Dunning, *J. Chem. Phys.* **90**, 1007 (1989).
125. (c) of Reference [34].
126. M.R. Pastrana, L.A.M. Quintales, J. Brandao, and A.J.C. Varandas, *J. Phys. Chem.* **94**, 8073 (1990).
127. O.I. Tolstikhin, S. Watanabe, and M. Matsuzawa, *Phys. Rev. Lett.* **74**, 3573 (1995).
128. O.I. Tolstikhin and H. Nakamura, *J. Chem.Phys.* **108**, 8899 (1998).
129. F.T. Smith, *J. Chem. Phys.* **31**, 1352 (1959).
130. O.I. Tolstikhin, S. Watanabe, and M. Matsuzawa, *J. Phys.* **B29**, L389 (1996).
131. J.C. Light and Z. Bacic, *J. Chem. Phys.* **87**, 4008 (1987).
132. G.V. Mil'nikov and H. Nakamura, *Comput. Phys. Commun.* **140**, 381 (2001).
133. V.A. Mandelshtam, T.P. Grozdanov, and H.S. Taylor, *J. Chem. Phys.* **103**, 10074 (1995).
134. W.H. Press, S.A. Teukolsky, W.T. Vetterling, and B.P. Flannery, *Fortran Numerical Recipies* (Cambridge University Press, Cambridge, 1996).
135. W.C. Gardiner, Jr. (ed.), *Combustion Chemistry* (Springer, New York, 1984).
136. H. Okabe, *Photochemistry in Small Molecules* (Wiley, New York, 1978).
137. M. Dupuis and J.J. Wendoloski, *J. Chem. Phys.* **80**, 5696 (1984).
138. S.L. Boyd, R.J. Boyd, and C.L.R. Barclay, *J. Am. Chem. Soc.* **112**, 5724 (1990).
139. J.H. Wang, H.C. Chang, and Y.T. Chen, *Chem. Phys.* **206**, 43 (1996).
140. K.A. Peterson and T.H. Dunning, Jr., *J. Chem. Phys.* **106**, 4119 (1997).
141. S.A. Perera, L.M. Salemi, and R.J. Bartlett, *J. Chem. Phys.* **106**, 4061 (1997).
142. E.L. Cochran, F.J. Adrian, and V.A. Bowers, *J. Chem. Phys.* **40**, 213 (1964).
143. P.H. Kasai, *J. Am. Chem. Soc.* **94**, 5950 (1972).
144. R.W. Fessenden and R.H. Schuler, *J. Chem. Phys.* **39**, 2147 (1963).
145. E. Kim and S. Yamamoto, *J. Chem. Phys.* **116**, 10713 (2002).
146. K. Tonokura, S. Marui, and M. Koshi, *Chem. Phys. Lett.* **313**, 771 (1999).
147. A. Fahr, P. Hassanzadeh, and D.B. Atkinson, *Chem. Phys.* **236**, 43 (1998).
148. H. Kanamori, Y. Endo, and E. Hirota, *J. Chem. Phys.* **92**, 197 (1990).
149. K. Tanaka, M. Toshimitsu, K. Harada, and T. Tanaka, *J. Chem. Phys.* **120**, 3604 (2004).
150. G.V. Mil'nikov, T. Ishida, and H. Nakamura, *J. Phys. Chem.* **A110**, 5430 (2006).
151. C. Moller and M.S. Plesset, *Phys. Rev.* **46**, 618 (1934); M. Head-Gordon, J.A. Pople, and M. Frisch, *Chem. Phys. Lett.* **153**, 503 (1988).
152. W.J. Hehre, R. Ditchfield, and J.A. Pople, *J. Chem. Phys.* **56**, 2257 (1972).
153. J.A. Pople, M. Head-Gordon, and K. Raghavachari, *J. Chem. Phys.* **87**, 5968 (1987).
154. M.J. Frisch, G.W. Trucks, H.B. Schlegel, G.E. Scuseria, M.A. Robb, J.R. Cheeseman, J.A. Montgomery, Jr., T. Vreven, K.N. Kudin, J.C. Brant, J.M. Millam, S.S. Iyengar, J. Tomasi, V. Barone, B. Mennucci, M. Cossi, G. Scalmani, N. Rega, G.A. Peterson, H. Nakatsuji, M. Hada, M. Ehara, K. Toyota, R. Fukuda, J. Hasegawa, M. Ishida, T. Nakajima, Y. Honda, O. Kitao, H. Nakai, M. Klene, X. Li, J.E. Knox, H.P. Hratchian, J.B. Cross, V. Bakken, C. Adamo, J. Jaramillo, R. Gomperts, R.E. Stratmann, O. Yazeyev, A.J. Austin, R. Cammi, C. Pomelli, J.W. Ochterski, P.Y. Ayala, K. Morokuma,

G.A. Voth, P. Salvador, J.J. Dannnenberg, V.G. Zkrzewski, S. Dapprich, A.D. Daniels, M.C. Strain, O. Farkas, D.K. Malick, A.D. Rabuck, K. Raghavachari, J.B. Foresman, J.V. Ortiz, Q. Cui, A.G. Baboul, S. Clifford, J. Cioslowski, B. Stefanov, G. Liu, A. Liashenko, P. Piskorz, I. Komaromi, R.L. Martin, D.J. Fox, T. Keith, M.A. Al-Laham, C.Y. Peng, A. Nanayakkara, M. Challacombe, P.M.W. Gill, B. Johnson, W. Chen, M.W. Wong, C. Gonzalez, and J.A. Pople, GAUSSIAN03, revision B.05, Gaussian Inc., Wallingford, CT (2004).
155. W.F. Rowe, R.W. Duerst, and E.B. Wilson, *J. Am. Chem. Soc.* **98**, 4021 (1976).
156. S.L. Baughcum, R.W. Duerst, W.F. Rowe, Z. Smith, and E.B. Wilson, *J. Am. Chem. Soc.* **103**, 6296 (1981).
157. S.L. Baughcum, Z. Smith, E.B. Wilson, and R.W. Duerst, *J. Am. Chem. Soc.* **106**, 2260 (1984).
158. P. Turner, S.L. Baughcum, S.L. Coy, and Z. Smith, *J. Am. Chem. Soc.* **106**, 2265 (1984).
159. D.W. Firth, K. Beyer, M.A. Dvorak, S.W. Reeve, A. Grushow, and K.R. Leopold, *J. Chem. Phys.* **94**, 1812 (1991).
160. T. Baba, T. Tanaka, I. Morino, K.M.T. Yamada, and K. Tanaka, *J. Chem. Phys.* **110**, 4131 (1999).
161. V.A. Benderskii, E.V. Vetoshkin, I.S. Irgibaeva, and H.P. Trommsdorff, *Chem. Phys.* **262**, 393 (2000).
162. T. Carrington, Jr. and W.H. Miller, *J. Chem. Phys.* **81**, 3942 (1984); **84**, 4364 (1986).
163. N. Shida, P.F. Barbara, and J. Almlöf, *J. Chem. Phys.* **91**, 4061 (1989).
164. N. Shida, J. Almlöf, and P.F. Barbara, *J. Phys. Chem.* **95**, 10457 (1991).
165. T.D. Sewell, Y. Guo, and D.L. Thompson, *J. Chem. Phys.* **103**, 8557 (1995).
166. K. Yagi, T. Taketsugu, and K. Hirao, *J. Chem. Phys.* **115**, 10647 (2001).
167. D. Babić, S.D. Bosanac, and N. Došlić, *Chem. Phys. Lett.* **358**, 337 (2002).
168. N. Došlić and O. Kḧn, *Z. Phys. Chem.* **217**, 1507 (2003).
169. R. Meyer and Tae-Kyu Ha, *Mol. Phys.* **101**, 3263 (2003).
170. M.D. Coutinho-Neto, A. Viel, and U. Manthe, *J. Chem. Phys.* **121**, 9207 (2004).
171. K. Yagi, T. Taketsugu, K. Hirao, and M.S. Gordon, *J. Chem. Phys.* **113**, 1005 (2000).
172. M.A. Collins, in *New Method in Computational Quantum Mechanics* edited by I. Prigogine and S.A. Rice (Wiley, New York, 1996), p. 389.
173. T. Takata, T. Taketsugu, K. Hirao, and M.S. Gordon, *J. Chem. Phys.* **109**, 4281 (1998).
174. K.C. Thompson, M.J.T. Jordan, and M.A. Collins, *J. Chem. Phys.* **108**, 564 (1998).
175. T. Taketsugu, N. Watanabe, and K. Hirao, *J. Chem. Phys.* **111**, 3410 (1999).
176. K. Yagi, T. Taketsugu, and K. Hirao, *J. Chem. Phys.* **116**, 3963 (2002).
177. K. Yagi, C. Oyanagi, T. Taketsugu, and K. Hirao, *J. Chem. Phys.* **118**, 1653 (2003).
178. Y. Guo, T.D. Sewell, and D.L. Thompson, *Chem. Phys. Lett.* **224**, 470 (1994).
179. R.J. Bartlett, *Annu. Rev. Phys. Chem.* **32**, 359 (1981).
180. R.A. Kendall, T.H. Dunning, Jr., and R.J. Harrison, *J. Chem. Phys.* **96**, 6796 (1992).
181. M.J. Frisch et al. Computer Code GAUSSIAN-98, Revision A.5, Gaussian Inc., Pittsburgh, PA (1998).
182. MOLPRO, a package of *ab initio* programs designed by H.-J. Werner and P.J. Knowles, Version 2002.1.
183. K. Yagi, G.V. Mil'nikov, T. Taketsugu, K. Hirao, and H. Nakamura, *Chem. Phys. Lett.* **397**, 435 (2004).
184. G.V. Mil'nikov, O. Kühn, and H. Nakamura, *J. Chem. Phys.* **123**, 074308 (2005).
185. Y. Marechal and A. Witkowski, *J. Chem. Phys. Phys.* **48**, 3697 (1968).
186. M.V. Vener, O. Kühn, and J.M. Bowman, *Chem. Phys. Lett.* **349**, 562 (2001).
187. C. Emmeluth, M.A. Suhm, and D. Luckhaus, *J. Chem. Phys.* **118**, 2242 (2003).
188. K. Heyne, N. Huse, J. Dreyer, E.T.J. Nibbering, T. Elsaessar, and S. Mukamel, *J. Chem. Phys.* **121**, 902 (2004).

189. A.Stökli, B.H. Meier, R. Kreis, R. Meyer, and R.R. Ernst, *J. Chem. Phys.* **93**, 1502 (1990).
190. Q. Xue, A.J. Horsewill, M.R. Johnson, and H.P. Trommsdorff, *J. Chem. Phys.* **120**, 11107 (2004).
191. S. Hayashi, J. Umemura, S. Kato, and K. Morokuma, *J. Phys. Chem.* **88**, 1330 (1984).
192. Y.-T. Chang, Y. Yamaguchi, W.H. Miller, and H.F. Schaefer III, *J. Am. Chem. Soc.* **109**, 7245 (1987).
193. B.S. Jursic, *J. Mol. Struct.: THEOCHEM* **417**, 89 (1997).
194. Y. Kim, *J. Am. Chem. Soc.* **118**, 1522 (1996).
195. S. Miura, M.E. Tuckerman, and M.L. Klein, *J. Chem. Phys.* **109**, 5290 (1998).
196. H. Ushiyama and K. Takatsuka, *J. Chem. Phys.* **115**, 5903 (2001).
197. K. Wolf, A. Simperler, and W. Mikenda, *Monatsc. Chem.* **130**, 1031 (1999).
198. N. Shida, P.F. Barbara, and J. Almlöf, *J. Chem. Phys.* **94**, 3633 (1991).
199. A. Garg, *Am. J. Phys.* **68**, 430 (2000).
200. C.S. Tautermann, A.F. Voegele, and K.R. Liedl, *J. Chem. Phys.* **120**, 631 (2004) (Third reference of Reference [42]).
201. F. Madeja and M. Havenith, *J. Chem. Phys.* **117**, 7162 (2002).
202. Z. Smedarchina, A. Fernandez-Ramos, and W. Siebrand, *Chem. Phys. Lett.* **395**, 339 (2004) (7th article in Reference [41]).
203. G.V. Mil'nikov, O. Kühn, and H. Nakamura, *J. Chem. Phys.* **123**, 074308 (2005).
204. C.S. Tautermann, M.F. Loferer, A.F. Voegele, and K.R. Liedel, *J. Chem. Phys.* **120**, 11650 (2004).
205. S. Shulman, *Techniques and Applications of Path Integration* (Wiley, New York, 1981), p. 284.
206. W.H. Miller, *Adv. Chem. Phys.* **XXV**, 69 (1974); ibid., **XXX**, 77 (1975).
207. T.J. Stuchi and R. Vieira-Martins, *Phys. Lett.* **A201**, 179 (1995).
208. H. Ushiyama, A. Arasaki, and K. Takatsuka, *Chem. Phys. Lett.* **346**, 169 (2001).
209. F.O. Ellison, *J. Am. Chem. Soc.* **85**,3540 (1963).
210. A.J.C. Varandas and J.D. Silva, *J. Chem. Soc. Faraday II* **82**, 593 (1986).
211. K.P. Kirby and E.P. VanDishoek, *Adv. At. Mol. Phys.* **25**, 437 (1988).
212. A. Bjerre and E.E. Nikitin, *Chem. Phys. Lett.* **1**, 179 (1967).
213. S. Chapman, *Adv. Chem. Phys.* **82**, 423 (1992).
214. K.G. Kay, *Chem. Phys.* **322**, 3 (2006); *J. Chem. Phys.* **100**, 4377 (1994).
215. K.G. Kay, *J. Chem. Phys.* **107**, 2313 (1997).
216. F. Grossmann, *Phys. Rev. Lett.* **85**, 903 (2000).
217. D.H. Zhang and E. Pollak, *Phys. Rev. Lett.* **93**, 140401 (2004).
218. M. Saltzer and J. Ankerhold, *Phys. Rev.* **A68**, 042108 (2003).
219. H. Eyring, *J. Chem. Phys.* **3**, 107 (1935); *Trans. Faraday Soc.* **34**, 41 (1938).
220. E. Wigner, *Trans. Faraday Soc.* **34**, 29 (1938).
221. P. Hänggi, P. Taikner, and M. Borkovec, *Rev. Mod. Phys.* **62**, 250 (1990).
222. W.H. Miller, in *Dynamics of Molecules and Chemical Reactions* edited by W.E. Wyatt and J.Z.H. Zhang (Marcel Dekker, New York, 1996), chap. 10.
223. D.G. Truhlar, B.C. Garrett and S.J. Klippenstein, *J. Phys. Chem.* **100**, 12771 (1996).
224. J. Cao and G.A. Voth, *J. Chem. Phys.* **105**, 6856 (1996).
225. M. Kryvohuz, *J. Chem. Phys.* **134**, 114103 (2011).
226. W.H. Miller, Y. Zhao, M. Ceotto, and S. Yang, *J. Chem. Phys.* **119**, 1329 (2003); C. Venkataraman and W.H. Miller, *J. Phys. Chem.* **A108**, 3035 (2004).
227. Y. Zhao and W. Wang, *Adv. Phys. Chem.* **2012** 483504 (2012).
228. I. Affleck, *Phys. Rev. Lett.* **46**, 388 (1981).
229. V.A. Benderskii, V.I. Gordanskii, and D.E. Makarov, *Chem. Phys.* **154**, 407 (1990).
230. V.A. Benderskii and D.E. Makarov, *Phys. Lett. A* **161**, 535 (1992).

231. E.T. Whittaker, *A Treatise on the Analytical Dynamics of Particles and Rigid Bodies* (Cambridge University Press, Cambridge, 1965).
232. V.A. Benderskii, P.G. Grinevich, D.E. Makarov, and D.L. Pastur, *Chem. Phys.* **161**, 51 (1992).
233. V.A. Benderskii, D.E. Makarov, and P.G. Grinevich, *Chem. Phys.* **170**, 275 (1993).
234. R.A. Marcus and N. Sutin, *Biochim. Biophys. Acta* **811**, 265 (1985).
235. M. Bixon and J. Jortner, *Adv. Chem. Phys.* **106**, 35 (1999).
236. A.V.Z. Brazykin, P.A. Frantsuzov, K. Seki, and M. Tachiya, *Adv. Chem. Phys.* **123**, 511 (2002).
237. J. Cao, C. Minichino, and G.A. Voth, *J. Chem. Phys.* **103**, 1391 (1995).
238. P. Pechukas, *Phys. Rev.* **181**, 166, 174 (1969).
239. Y. Okuno and S. Mashiko, *Int. J. Quant. Chem.* **102**, 8 (2005).
240. A.J. Marks and D.L. Thompson, *J. Chem. Phys.* **96**, 1911 (1992).
241. A.J. Marks, *J. Chem. Phys.* **114**, 1700 (2001).
242. M.S. Topaler and D.G. Truhlar, *J. Chem. Phys.* **107**, 392 (1997).
243. Y. Zhao, G. Mil'nikov, and H. Nakamura, *J. Chem. Phys.* **121**, 8854 (2004).
244. W.H. Miller, S.D. Schwartz, and J.W. Tromp, *J. Chem. Phys.* **79**, 4889 (1983).
245. Y. Zhao, W. Liang, and H. Nakamura, *J. Phys. Chem.* **A110**, 8204 (2006); Y. Zhao and H. Nakamura, *J. Theo. Comp. Chem.* **5**, 299 (2007); Y. Zhao, M. Han, W. Liang, and H. Nakamura, *J. Phys. Chem.* **A111**, 2047 (2007).
246. S. Hammes-Schiffer and J.C. Tully, *J. Chem. Phys.* **103**, 8528 (1995).
247. B.G. Levine and T.J. Martinez, *Annu. Rev. Phys. Chem.* **58**, 613 (2007).
248. Y. Teranishi, H. Nakamura, and S.H. Lin, invited talk at TACC-2012 in *Theory and Applications of Computational Chemistry: The First Decade of the Second Millennium* pp. 119–130 (American Institute of Physics, 2012).
249. J.O. Richardson and S.C. Althorpe, *J. Chem. Phys.* **134**, 054109 (2011); J.O. Richardson, S.C. Althorpe, and D.J. Wales, *J. Chem. Phys.* **135**, 124109 (2011).
250. S. Takagi, *Macroscopic Quantum Tunneling* (Cambridge University Press, Cambridge, 2002).
251. E. Chudnovsky and J. Tejada, *Macroscopic Quantum Tunneling of the Magnetic Moment* (Cambridge University Press, Cambridge, 2005).

Index